The Buzzworm Magazine Guide to

ECOTRAVEL

100 unforgettable adventures throughout the world,
where the new ecotraveler will find the rewarding
delights of travel with the Earth in mind.

FROM THE EDITORS OF BUZZWORM MAGAZINE

BUZZWORM BOOKS
BOULDER, COLORADO

THE BUZZWORM MAGAZINE GUIDE TO ECOTRAVEL

Editor-in-Chief
Joseph E. Daniel

Editors
Ann Carey
Lisa Jones

Editorial Staff
Julie Carey
Mike Grudowski
Rebecca Hess
Deborah Houy
Ilana Kotin
Marina Lindsey
Mike Medler
Andrea Merson
Scott Woodford

Design and Production
Steve Harley
Karen Oldenburg
Christy Brennand
David McCloskey
Jennifer L. Wolcott
Viktor J. Strayer

BUZZWORM BOOKS
Founder & Publisher
Joseph E. Daniel

Co-founder
Peter Stainton

Cover Photo: Lanny Johnson/Mountain Stock

Copies of *Ecotravel* may be ordered directly by mail. Special pricing available for premium, gift and fundraising bulk orders. Call toll-free (800) 333-8857 for price and shipping information.

Distributed to the trade in the United States by Publishers Group West.

Printed by Arcata Graphics Company, El Segundo, California.

THE BUZZWORM MAGAZINE GUIDE TO ECOTRAVEL
ISBN 0-9603722-8-8

BUZZWORM BOOKS is an imprint of BUZZWORM Magazine

BUZZWORM, Inc.
2305 Canyon Blvd., Suite 206
Boulder, CO 80302
(303) 442-1969

Printed on recycled paper with 10% post-consumer waste

PRINTED IN THE UNITED STATES OF AMERICA
10 9 8 7 6 5 4 3 2 1

"Make voyages.
Attempt them.
There is nothing else."

—Tennessee Williams

TABLE OF CONTENTS

THE NATURE OF ECOTOURISM

100 UNFORGETTABLE ADVENTURES

RESOURCES

Masai women present their crafts and beadwork for sale to Western tourists in Kenya, East Africa.

ECOTOURISM: CLOSING THE GAP BETWEEN INTENT AND ACTION

By Lisa Jones

Last fall, I went to the Okavango delta—a pristine wilderness in the southern African nation of Botswana. I met a group of birders there; a friendly, lively group that was absolutely smitten with birds. When they asked me if I'd like to go with them and their private guide to visit a nearby bird rookery, I accepted instantly. They rattled off the names of species as our boat zoomed through the papyrus-lined channels; they were infectiously happy and excited as they pointed out the impalas on the banks and crocodiles in the water. Their boatman throttled back as we neared the rookery, where a marabou stork had just hatched three chicks. We putted up to the tree, cut the engine, drifted closer still and bumped soundly into the trunk of the tree that held the nest. Then the guide stood up and pulled down the branch supporting the jumble of twigs and grasses and the three disconcerted chicks, while the mother hopped away to a nearby tree. My birding companions stuck their video cameras and still cameras right into the family circle so they could get perfect photographs.

At dinner that night the guide bemoaned the decline in bird numbers at the rookery in recent years. He said it probably had something to do with its discovery by increasing numbers of tourists. I am sure that his concern was sincere, yet I'm absolutely sure it hadn't even occurred to him that nearly clambering in with the chicks that afternoon had made an impact.

Ecotourism is about many things, but in its broadest sense it is about closing the gap between our intentions and our actions toward the environment. To borrow from Henry David Thoreau, it's about traveling deliberately. The Ecotourism Society defines it as "responsible travel which conserves environments and sustains the well-being of local people." That means much more than carrying a camera rather than a gun. It also means more than stopping the boat before it crashes into a marabou stork's nest. It means generating revenues for the protection of the area through user fees and royalties. It also means injecting money into the local economy, so that local people consider the stork's preservation economically beneficial. Money generated by ecotourism carries with it a simple but powerful message—Nature Is Valuable. Unless local people come to view environmental conservation as economically profitable, we can kiss many of our unexploited wildlands goodbye.

Ecotourism is most easily defined in the developing world, where exploding populations and widespread poverty often force local residents to turn to natural areas in search of sustenance. There, ecotourism can take the pressure off the land by generating employment like guiding, offering lodging and more. "If people can't make a living from the forest, they'll start slashing and burning to plant crops," notes one conservationist who is helping to set up small ecotourism enterprises in Guatemala's lowland jungle.

In the developed world, where economies and population levels are generally more stable, tourism can help conservation by fostering environmental knowledge and activism. And wherever park fees and other royalties are applied to the management of the area rather than to the government's general fund, tourism revenues constitute a shot in the arm for management and protection.

But when we sign up for an outfitted trip, how do we know that our money will go toward protecting the place to which we travel?

Reading this book is one step. We've carefully questioned outfitters, nonprofit and for-profit alike, on how they benefit environmental conservation in the areas they visit. But it's only a step. Ecotourism is in its infancy, and while the following pages describe successful ecotourism projects, evidence of exactly how this industry helps environments and economies on a larger scale is spotty. It involves a network of relationships between cultures, landscapes and money as varied as the landscapes and cultures themselves. With ecotourism at this fledgling stage, the responsibility for defining what it is usually falls in the lap of the traveler.

Many environmentalists justly ask, "Why travel?" Overseas travel burns a hefty amount of fossil fuel, and there's no denying that tourists make an impact on the people and wildlife of the countries they visit.

Viewed from the balcony of the ivory tower, this is bad. When travelers are insensitive to the environments they visit, the results are indeed destructive. But when traveling is viewed from the ground, two hard facts become apparent: People who can afford to fly to faraway places will do so; and exploding populations and poverty are eating away at some of the most precious remaining wilderness on Earth. In many places, ecotourism presents as good an alternative as any to make an area support the people that depend upon it. Unlike mining, logging or agriculture, ecotourism's success depends upon the land retaining its natural features.

Ecotourism isn't going to save the world. It works only in places that people want to visit. And in order to avoid causing too much environmental impact, numbers of travelers must generally be limited, which puts a ceiling on how much employment it can generate.

But ecotourism is a tool. It certainly isn't the only tool we have for fashioning a better world, but it's a tool with a wonderful ability to create incentive to protect natural treasures. And in the absence of well-worn standards for what constitutes ecotourism in every corner of the world, today's ecotravelers are in a position of great responsibility and power.

Lisa Jones is Ecotravel Editor for BUZZWORM *Magazine.*

NATURE TOURISM IN KENYA: WHAT HAPPENED?

By Costas Christ

Kenya, with its enormous herds of wildlife and towering mountain backdrops, has a place in the imagination of most nature-loving travelers. But government corruption and widespread poverty combined to bring this beautiful country—and its once-prodigious wildlife herds—to its knees. Drastic measures such as shooting poachers have been applied to the grim situation in Kenya, and this once-mighty wildlife sanctuary is starting to rebound. Former Kenyan resident **Costas Christ** *reports:*

When a group of heavily armed poachers drove a pickup truck into Kenya's Meru National Park in October 1988 and in broad daylight slaughtered the last five surviving white rhinos in the country, it became unmistakably clear that wildlife conservation in Kenya was in crisis. In the late 1980s, some reports declared that an average of four elephants per day were being killed by poachers. In one particularly harrowing incident, a group of tourists witnessed a family of elephants being cut down with automatic rifles by poachers in northern Kenya.

What happened? Kenya had been a pioneer in developing nature tourism in the 1960s and 1970s. Nature lovers flocked to this wonderful country to see its spectacular wildlife, magnificent mountains and pristine beaches. When hunting was officially banned in Kenya in 1977, due to public pressure to protect its wildlife heritage, tour operators further diversified their wide range of safari offerings. At a time when the word "ecotourism" had yet to be uttered, nature tourism in Kenya proved to be a big success: Travelers came armed with cameras, not guns, foreign-dominated tour companies and hotels were replaced by locally owned businesses and the management of certain parks and reserves, most notably Masai Mara and Amboseli, was placed under the control of indigenous people in the area. In 1988, tourism had replaced coffee and tea exports to become Kenya's number-one source of foreign exchange earnings.

As tourism revenues increased, however, so did corruption and mismanagement in the Kenyan government, whose responsibility it was to preserve Kenya's wildlife and to care for the majority of the country's national parks and reserves. When Kenya's last surviving white rhinos were butchered in Meru National Park, critics were quick to ask how poachers could drive into a national park in daylight, shoot each of the five animals (which were under 24-hour armed guard by park rangers), use chainsaws to cut off their horns and then drive out of the park without getting caught by the authorities, who launched an immediate "manhunt" for the criminals responsible.

Efforts to stop poaching were further crippled by lack of funding: There wasn't enough equipment for park rangers, enough fuel for vehicles or enough money for administration. Game rangers armed with World War II rifles were pitted against poachers with AK-47s. Morale was rock-bottom. In addition, uncontrolled tourism in some parks and reserves was leading to overdevelopment of lodges and hotels within the borders of protected areas, as well as destruction of vegetation and the life cycles of wild animals due to too many vehicles driving off of designated roadways.

As its wildlife populations dwindled, the Kenyan government found itself on the defensive both domestically and internationally. Environmental groups were reluctant to offer support to Kenya's conservation efforts with such dismal results on the record. Kenya took drastic action in 1989, and the results have been impressive.

The most significant step taken was to move the management of the parks out of the hands of government bureaucrats and into the newly established Kenya Wildlife Service, currently headed by Dr. Richard Leakey. Under the Kenya Wildlife Service, the management of parks and reserves was put under nongovernmental, autonomous control. More money was plowed directly back into the various protected areas and their surrounding communities. This has also given a clear incentive to each park and reserve to manage itself more efficiently in order to generate more income locally, leading to more jobs and better working conditions for park staff. In its most dramatic stance to date, it equipped and authorized park rangers to shoot poachers on sight. While the final success of the Kenya Wildlife Service will be determined in the years to come, one result of its efforts has been immediate—large-scale poaching in Kenya has come to a virtual halt.

Alarmed by research reports that elephants could become extinct from the wilds of East Africa in as few as 10 years, President Daniel arap Moi of Kenya also took action, making Kenya the first country in Africa to declare a complete stop to ivory exports. He then led an international crusade to ban the sale of ivory worldwide. This was most dramatically publicized in his public burning of 12 tons of ivory worth $3 million in 1989. In October of that year, Kenya successfully led a number of the signatory nations of the Convention on International Trade in Endangered Species (including the United States) to support a total ban on any commerce whatsoever in elephant products.

While this strong stance proved to be controversial and led to serious tensions between Kenya and its neighbors to the south, Zimbabwe and Botswana, both of which favored continuation of the legal ivory trade, there is no question that in Kenya, at least, the result has been a dramatic decrease in elephant poaching. While the Kenyan government deserves credit for its recent success in combatting poaching, it could well spend more energy on sporadic but continued attacks upon tourists themselves.

Kenya still remains one of the world's premier destinations for eco-tourism. It has the experience and now the renewed commitment to make sure that its priceless natural treasures remain intact.

Costas Christ is a regional coordinator for the Ecotourism Society and owner of Tamu Safaris, a small ecotour operator specializing in Africa and based in West Chesterfield, New Hampshire.

ADVENTURING, ADVOCACY AND LIFE CHANGE ON THE GRAND CANYON

By Rob Elliott

*The Grand Canyon was a stunning enough sight to send its early explorers such as John Wesley Powell into rhapsodies. It remains a transformative experience for many of the 22,000 people who float through it on the Colorado River each year. Its power to inspire its visitors played a major part in an upswell of activism that successfully pressured the federal government to take steps to protect it. River runner and environmental activist **Rob Elliott** reports:*

One woman wrote to me after rafting the Colorado River through the Grand Canyon: "How could anyone make the canyon voyage and emerge unchanged? There is a quality about the river and the desert wilderness that mirrors each traveler's life in its barest and purest essentials, challenging us to separate the important from the unimportant."

Many people have shared similar comments with me. Although I have some great adventures still ahead, I look back over the 28 years I've accompanied people on their journeys through the Grand Canyon with a sense of reverence—how blessed I've been to watch people react to the river. The river hasn't only been an agent of personal change, it has helped catalyze one of the most hopeful political changes I've seen in a decade.

After witnessing the flooding of Glen Canyon beneath Lake Powell in southern Utah in the early 1960s, I vowed to do whatever possible to keep dams out of the Grand Canyon. I didn't realize then that over the next 28 years the riparian zone through the Grand Canyon would come under heavy, persistent attack, day in and day out, from the flows released by Glen Canyon

Dam to generate electricity. Until recently these flows fluctuated up to 13 feet daily, saturating the beaches, dropping away in a few hours, and slumping tons of sand into a clear and hungry river. Over the intervening decades, we lost up to 50 percent of the sand beaches and three of four endangered fish species from the Colorado River through Grand Canyon. By 1989, the cumulative evidence of significant environmental degradation had become compelling.

The first step toward protecting the Grand Canyon was to force the Department of Interior to commit to conducting an Environmental Impact Statement (EIS) on the operations of Glen Canyon Dam, something the department had been resisting for over a decade. America Outdoors (a national organization of 200 whitewater rafting outfitters) joined a broad coalition of seven national and regional environmental groups, and we went looking for people who cared enough to get involved. People who had journeyed through the Grand Canyon seemed a natural group to ask for help. We figured these people were often well-connected, articulate and, most importantly, gave a damn.

In 1988 and 1989, 8,000 people, many of them veterans of river trips, wrote members of Congress imploring them to "please help the Grand Canyon." It worked. Politicians leaned on the Secretary of Interior and in July 1989 he agreed "to do the right thing" and conduct an EIS on the operations of Glen Canyon Dam. Some 17,000 more people wrote to get on the EIS list of "interested public" to stay involved while the fluctuating flows raged on.

The next step was to force the secretary to implement interim operations that would hold the fluctuating flows in check until the completion of the EIS. There was only one way: by pressing for passage of the Grand Canyon Protection Act first introduced by Arizona Republican Senator John McCain in 1990. Although the act has met several hurdles and may or may not pass, it forced the secretary, against considerable opposition by public utilities, to implement interim protective flows until permanent protection can be instated.

The Grand Canyon Protection Act got as far as it did because Congress and Senator McCain received a flood of letters backing the act. The letters were written by people who cared deeply about the Grand Canyon, many because the Canyon had given them something—a perspective, a prod into looking at what's important in life and what isn't.

If there's a story about traveling down the Grand Canyon, it is this: Beware. Given the right ingredients and inclination, enough time and geography, you might go back home and take on a few new hobbies such as writing letters to Congress, helping organize a community recycling program, lobbying to preserve a wetland in the next county or maybe even throwing in the towel and doing something entirely different with your life. But don't go looking for the life change. Go adventuring, go play and the life change may come looking for you.

Rob Elliott is a second-generation river runner who owns Arizona Raft Adventures in Flagstaff, Arizona. He is the Conservation Chairman of America Outdoors and a member of the Board of Trustees of the Grand Canyon Trust.

ZIMBABWE'S CAMPFIRE PROGRAM

By Kim Larsen

If cash-poor rural communities see wildlife as an obstacle to their survival, they won't likely want to conserve it. But if wildlife becomes their cash crop, they will actively defend it. The CAMPFIRE project funnels money generated from trophy hunting and disperses it among local rural communities. This provides the means for community development, encourages locals to get involved in professional wildlife management and guiding, and goes a long way toward changing negative attitudes about wildlife among rural people. Writer **Kim Larsen** *reports:*

One of the most heartbreaking sights I came across in Zimbabwe was a fence that a farmer had rigged in an attempt to keep elephants off his farm. In sections the fence was a woven thicket of gathered brambles and thornbush, but connecting these sections were stretches of twine with objects hanging from it—rusty tin plates, scraps of clothing, the odd plastic jug (there isn't much refuse in the bush)—anything that might frighten the elephant off its course. The fence was as high as my knee, which means about mid-shin to the average elephant. Without so much as a blink, the beast would lumber right through this ephemeral barrier. The farmer had to know this, but for him it was either build it or do nothing, and that was unacceptable. His farm—an acre or two of undernourished maize—had been ravaged only several nights before by a family of elephants. Their footprints were still fresh—a chaos of blasted pits crawling with termites. What the elephants had not eaten they had crushed. The crop was in ruin. The farmer had done what he could to protect his future harvests—erect a pitifully useless fence.

There is a program in Zimbabwe that would reshuffle the deck for this farmer. Placing the stewardship, management and profit incentive of wildlife and other resources firmly in the hands of local inhabitants, CAMPFIRE (Communal Areas Management Program for Indigenous Resources) represents an approach to rural development that unambiguously stresses the necessity for ground-level participation.

For decades, the conflicting needs of people and wildlife have polarized efforts to aid in their survival. A solution that would benefit both seemed permanently locked away in wishful thinking. CAMPFIRE cracks the safe and delivers a reality-based approach. Not only does it address the needs of humans and animals alike, but it makes the two camps indispensable to one another.

However, the reality on which CAMPFIRE is based is one that is hard for elephant lovers unfamiliar with the situation on the ground to accept: Trophy hunting is the key to the program's success. CAMPFIRE sets up arrangements between local councils and safari concessions catering to international game hunters, in which the lion's share of the proceeds returns to the "producer communities." Out of an estimated total in Zimbabwe of 77,000 elephants, 0.7 percent, all adult males, are taken as sport hunting trophies annually. This off-take is of little consequence to an elephant population that has a growth rate of 5 percent a year—but it's enormously consequential to those who reap the economic benefits of trophy hunting. Previously, the people who shared the elephants' habitat had no reason to view these (and other) animals as anything more than a crop-destroying, life-threatening, government-protected scourge. Now, with the CAMPFIRE initiative, wildlife becomes their cash cow. In places where CAMPFIRE projects have matured, the beneficiaries say so themselves: "Wildlife are our cattle." In light of the exalted position of domestic livestock in African culture, that's saying a lot.

What's new here is not the concept of the sustainable use of wildlife—this has been practiced by private landholders since Zimbabwe passed the 1975 Parks and Wildlife Act, resulting in an increasing acreage of privately held land devoted to wildlife and a heightened awareness of its money-making potential. What is new about the CAMPFIRE project is that small rural communities occupying communal lands are in charge. They manage and benefit from the wildlife at their disposal. It is precisely through this kind of participation at the least visible levels of society that CAMPFIRE realizes its goal: Only when the economically dispossessed recognize that wildlife is an asset will they want to preserve it. Where CAMPFIRE is effective, poachers will be shunned because they no longer provide a social service but rather rob the people of their riches. Cattle will not be pursued as the ultimate possession or commodity, thus yielding grazing lands back to wildlife. Subsistence agriculture, which in these arid regions is unsound both economically and ecologically, will also give way to wildlife.

When a single elephant generates 11,500 Zimbabwean dollars (there are currently $5.2 Zim to $1 US; for years there were $2 Zim to $1 US) for a community in a country where the average per capita income is approximately $3,640 Zim, heads turn and behavior changes. Once that money is delivered, the local wildlife committee will no longer be "out with their dogs," i.e., poaching, during planning meetings. Such signs of skepticism are melting away as CAMPFIRE comes into its own. The biggest problem the Shangaan-speaking villagers of Chikwarakwara had to face when they were informed at a council meeting that elephants alone would yield them $34,500 Zim for 1991 was this: Their language has no word to express such a high number.

When the money came in for Chikwarakwara, the council made the decision to distribute it to the villagers at a ceremony that would concretize their earnings. Sixty thousand dollars was procured in $20 notes and ushered in to the proceedings in a wire basket held high above the chairman's head. Schoolchildren sang their specially composed song, "Wildlife Is Wonderful," and villagers were summoned one by one to receive their allotted $400. On the spot, they took half of this amount and deposited it into a plastic bucket for community projects—which, after much deliberation, had been decided in advance would go toward building a new school and buying a grinding mill. Each household brought home $200, a prodigious sum for these people.

None of this would be possible without big game hunters from the developed world finding their way to the Zimbabwean bush with money and leisure to burn. CAMPFIRE is also working on projects that would derive their income from non-hunting tourist activities (photo safaris, horse treks, canoeing, etc.), but these require a more complicated infrastructure that will take more time to bring up to speed. Meanwhile, I think of that farmer behind his strung junk, working desperately to salvage what he could of his crop, and hope that soon he'll have cause to trample the fence himself and invite the elephants in gladly.

For more information on trophy hunting safaris that benefit CAMPFIRE, contact CAMPFIRE, P.O. Box 4027, Harare, Zimbabwe, Africa, or National Parks and Wildlife Management, Hunting Department, P.O. Box 8365, Harare, Zimbabwe, Africa. For more information on CAMP-FIRE in general, contact the Zimbabwe Trust's US-based office at 1401 16th St. NW, Washington, DC 20036, (202) 939-9655.

Kim Larsen is a writer based in New York City.

WHERE THE MONEY GOES
by Kreg Lindberg

In the developed world, many of the costs of environmental protection are hidden. For instance, most of the National Park Service's budget comes from the government's general fund, not from entry fees to the parks. But when it comes to traveling, assumptions like these are the first things that should be left at home. Paying fees in foreign national parks or other attractions will fund their protection; flying in on the national airline will funnel money into the country's economy. Ecotourism economist **Kreg Lindberg** *writes on spending thoughtfully:*

Think back to your last vacation. Let's say you took an "ecotour" to Indonesia, stopping first for some snorkeling in the Bunaken Marine Park off the northern coast of Sulawesi. You were awestruck by the variety of coral and colorful fish. You were able to tell your friends not only of your chance encounter with a sea turtle, but also of the eerie feeling that crept into your stomach as you swam over the deep blue 1,000-foot drop-off on the edge of the reef. All in all, you were quite impressed, but you were disappointed when you encountered Bintang beer bottles (the local brew) nestled in the coral and when your boat operators dropped the anchor into the reef, destroying fragile (and slow-growing) coral in the process.

After a few days of exploring Bunaken, you flew off to wild Kalimantan (the Indonesian portion of Borneo) in search of the aptly named proboscis monkey and the elusive orangutan. Your trip to the tropical rainforest of

Gunung Palung National Park was a success; you woke early in the morning to see the proboscis along the riverbanks, while by good fortune you were able to see the orangutan, the "man of the forest," as he is known in Indonesia. Again, you were impressed by the trip, but disappointed when you came across the black thread nets used by local residents to catch exotic birds for sale in the markets of Singapore, Jakarta (the capital of Indonesia) and elsewhere.

As you completed your vacation, relaxing in magical Bali, you reflected on your trip and what effect it might have had in light of statements by your tour operator and the press that ecotourism does wonders for conservation and rural development. In short, you questioned whether your travels had supported "sustainable development."

According to the Ecotourism Society, a working definition of ecotourism is "responsible travel which conserves environments and sustains the well-being of local people." How might your trip achieve these goals? In terms of economic effects, there are two broad avenues through which ecotourism can contribute to conservation and the well-being of local people. First, tourism dollars can support the establishment and improved management of conservation areas. Second, tourism dollars can provide much-needed revenues for remote communities, thereby improving living standards and simultaneously reducing pressure to degrade the natural environment.

Perhaps the most obvious way tourism supports conservation is through the payment of entrance fees or other fees levied on tourist activities (such as hotel and departure taxes). When these fees go back to the parks department or whatever organization owns the ecotourism site, they can be used to finance the purchase and management of conservation areas.

In several countries around the world, ecotourism entrance fees are already used to finance private nature reserves or large portions of government wildlife-conservation budgets, and this trend is expected to increase in the future. There is thus a direct link between tourism and conservation.

Remember your snorkeling trip at Bunaken? There was no entrance fee, and government funding for park management is minimal. There is simply not enough money to hire personnel to pick up the beer bottles and other litter. There is no money to train park personnel on the impacts of activities such as dropping anchors on the reef. There is no money even to purchase a boat for park rangers. They have to ride on tour boats, and you can hardly expect them to regulate the industry that provides their only means of transportation. An entrance fee is being considered; if implemented it may be crucial in improving management of the park.

Some tour operators have opposed significant entrance fees in the belief that higher costs will reduce the number of consumers. As a consumer you should make it clear that you are willing to pay these fees. They may seem high if you are used to free or low-cost entrance to parks in the US and elsewhere. In addition, you may be disappointed insofar as these fees don't "buy" you much in the way of infrastructure or services.

Keep in mind, however, that in the US we pay for most park management through our taxes. We can hardly expect Indonesians to pay, through taxes, so that we can visit their parks at lower cost to ourselves. In addition, much of the cost of ecotourism attractions comes not from providing services or building infrastructure but from the "opportunity cost" of not using the areas for logging, agriculture or some other consumptive use. This hidden cost, often borne by local residents living at subsistence level, can be quite high. Furthermore, entrance fees are rarely higher than $10, which is a miniscule amount when you consider the thousands of dollars you spent for the whole trip.

In the more common case where entrance fees are channeled into the central government treasury, tourism revenues will have a smaller impact. However, they may play an important role in increasing general political support for conservation. Cash-strapped governments in both developing and industrialized countries often make decisions based on what alternative brings in the most money to the national treasury or the economy in general. Tourism is a way to make conservation pay, and thus a way to make conservation happen. As they say in East Africa, "Wildlife pays so wildlife stays."

These indirect impacts arise whenever your spending benefits the host country. Flying on the national airline, buying local products rather than imports, and using local guides and operators all contribute. For example, your flight on Garuda Indonesia rather than United Airlines earned the country over $1,000 (and the food was delicious!).

Kreg Lindberg is a research associate with the Ecotourism Society and a Ph.D. candidate in economics and tourism at Oregon State University. He recently authored the report "Policies for Maximizing Nature Tourism's Ecological and Economic Benefits" for the World Resources Institute in Washington, D.C. See page 223 for ordering information.

ECOTOURISM COMES TO SHANGRI-LA: TREKKING IN NEPAL

By Stan Stevens

*Nepal has long been beloved by trekkers and mountaineers, who have loved parts of this remote and fragile land nearly to death. Visitors have left piles of garbage at the feet of their favorite mountains and taken showers warmed with wood that this deforested land can ill afford to lose. There's good news, too. Away from the beaten path, the impacts are surprisingly light, while the two most heavily traveled areas in the country are the focus of initiatives to find alternative heating sources and to clean up garbage. Geographer **Stan Stevens** reports:*

Twenty years ago scarcely a thousand trekkers per year visited the Mt. Everest area. Now 9,000 do. The Annapurna Range, Nepal's other major trekking destination, hosts three times as many tourists per year. All told, some 60,000 trekking permits are issued annually for journeys into the high country of this remote Himalayan kingdom. This boom in adventure travel has brought both economic opportunities and environmental pressures to these areas.

Travelers have brought prosperity to the Sherpas of the Mt. Everest region: Two-thirds of the families in the region have income from tourism and have successfully integrated subsistence agriculture and livestock raising with careers in trekking, mountaineering and operating lodges and shops. Some villagers in the Annapurna Range have also benefitted from trekking tourism, especially from opening tourist lodges.

Yet the environmental costs have been high. The influx of tourists has accelerated deforestation along heavily-traveled routes, where trees are used to construct lodges and to fuel fires for cooking. Tourism has also imported tons of garbage to Nepal, especially the infamous "garbage trail" to Mt. Everest. Litter has also become increasingly visible in the Annapurna Range.

But promising efforts are now under way in both the Everest and Annapurna regions to develop environmentally sound trekking and mountaineering tourism. The Everest region has been a national park since 1976 and a World Heritage Site since 1980. The Annapurna area is managed by the King Mahendra Trust for Nature Conservation, Nepal's leading nongovernmental conservation association, under its Annapurna Conservation Area Project (ACAP). Programs are now being implemented in both of these protected areas which seem likely to become models for ecotourism planning for mountain regions.

Steps toward ecotourism

In the upper Modi Khola valley in the Annapurna region, a kerosene depot was established to make fuel available to both trekkers and lodge owners. Local lodge owners agreed five years ago to halt their use of fuel wood. Kerosene is seen as a stopgap measure to take immediate pressure off the forests while small-scale hydroelectric projects are completed. ACAP recently completed its first hydroelectric plant in the area, and more are planned for the next few years. In the Everest area, hydroelectricity is likewise seen as the key to decreasing the use of fuel wood. Two small hydro facilities are already supplying cooking energy to some lodges in two heavily touristed sites, and more hydropower is planned. Because the turbines are powered by water diverted from small streams and returned downstream, there is little environmental impact. However, no alternative energy has been developed where it is needed the most: around the highest-altitude lodges. The harvesting of slow-growing high-altitude juniper for fuel wood continues unabated.

The garbage problem is also under attack. Lodge owners in the Modi Khola valley have built garbage pits at each lodge site, and the local high school now conducts an annual trail cleanup expedition at the end of tourist season. In the Everest region, a Sherpa Pollution Control Committee was founded during summer 1991 with funding from World Wildlife Fund. The committee has already led the first Sherpa cleanup of Everest base camp and established a series of garbage dumps. There are plans to develop a regional recycling program and a system that will require mountaineering expeditions to leave a substantial deposit against their cleanup of their base camps.

There is much interest in Nepal not only in making trekking and mountaineering tourism more environmentally sensitive, but also in using tourism as a resource for conservation and development. One of the fundamental principles of ACAP is that tourism itself will support the administration of the conservation area and its projects. ACAP will use trekker entrance fees to support village-initiated projects such as agricultural extension programs, village water-supply systems, hydroelectric power and fodder tree nurseries.

Stan Stevens is an assistant professor of geography at Louisiana State University and is the author of Claiming the High Ground: Sherpas, Subsistence, and Environmental Change in the Highest Himalaya (*University of California Press, 1993*).

QUESTIONS TO ASK OUTFITTERS

By Frances Gatz and Jerome Touval

When you're looking for an ecotourism outfitter, don't be dazzled by leafy green logos or natural-sounding names. The most reliable judge of a tour company's environmental credentials is you. If you ask the right questions before you even start to pack, you are more likely to have an environmentally sound, enjoyable trip. Ecotourism experts **Frances Gatz** *and* **Jerome Touval** *have some suggestions for what to keep in mind when you plan your trip:*

Whether you are planning a wilderness camping trip to Grand Teton National Park or taking a break from the sun on the Yucatan peninsula to visit Sian Ka'an Biosphere Reserve, the net effect is that nature-related travel accounts for billions of dollars in expenditures each year.

Who benefits from the cost of your trip?

Do local people participate in your selected travel programs? Are they rewarded for their involvement? Do tour operators employ local nature guides, use locally owned hotels and restaurants, and work with national airlines and tourist boards? Do the local profits of a travel program contribute to sustainable development in the areas that you visit? Do local people feel a greater pride in their environment as a result of the economic benefits from tourism?

Do tour operators evaluate and seek to reduce the environmental impacts of their programs?

Examine each company's brochures. Do they provide information that describes the need to protect delicate areas? Do their business practices support the conservation efforts of the countries that you visit? Are environmentally sensitive guidelines in place for both staff and clients?

Bear in mind that the tour operator that markets a travel program may not be the same company that conducts the trip locally. Knowing about the on-site operator is also critical. Ask questions to discern why the on-site operator was selected. Can you trust your tour operator to have made the best selections?

Do tour organizers evaluate and seek to reduce the environmental impact of the tourist lodges and transportation? What are their methods of waste disposal? Do they recycle where possible? Are trails and walkways in place where they will lessen tourism impacts?

The people who live in and near the resource will ultimately determine its future. While we think of pristine areas like tropical rainforests as belonging to all of humankind and voice our support for their preservation, we should also recognize the right of their residents to self-determination. Ecotourism can be a valuable tool for conservation because it can provide an environmentally sustainable way for local residents to earn a living.

In the best of all possible worlds, an ecotourism management plan developed by local residents, park managers, government planners, conservation groups and commercial entities gives guidance as an area gains popularity as a travel destination. Tour operators have the expertise and insight to make an enormous contribution to protected-area management planning.

What relationship do tour operators have with the local communities and the management in the protected areas that you visit?

Are they known to be cooperative and interested in the well-being of local people and the environment of the region? How do they see the quality of life of local people improved through tourism? Are people and destinations depicted respectfully and realistically in travel brochures? Do tour operators offer ways to maintain contact with the people that you meet and share your experiences at home?

Ecotourism encourages a respect for local cultures and reinforces a community's character and identity. The authenticity of local cultures is an integral part of what compels us to visit new and unusual places. While the world is being homogenized, we can still make a statement against the drift toward crushing sameness through our actions.

In a society driven by the forces of supply and demand, you and others like you can demand responsibility from the tour operators, hotels, restaurants and other services that you patronize on your trip. Who benefits from this? Local economies and the natural areas you visit benefit. Wildlife benefits from being guaranteed a place to reproduce and flourish. You and your traveling companions enjoy a unique experience. Perhaps most importantly, the biodiversity of our planet stands to make the biggest gains of all.

Dr. Frances Gatz is a co-organizer of the Ecotourism Society and Director of Environmental Workshops for International Expeditions in Helena, Alabama.

Jerome Touval worked for 11 years as a wildlife biologist for the US Fish and Wildlife Service and the US Bureau of Land Management. He is currently with the Conservation Biology Program at the University of Maryland.

100 UNFORGETTABLE ADVENTURES

Great low-impact trips, from your backyard
to the edge of the Earth.

THE ALASKA WILDLAND SAFARI
Explore Kenai and Denali with Alaska Wildland Adventures

Who says you have to suffer to have an exciting wilderness experience? Alaska Wildland Adventures provides trips that are ideal for families, outdoor enthusiasts and anyone else wishing for more intimate encounters with nature, without stepping beyond the threshold of comfort. Alaska Wildland Safari features hikes in mountain meadows, a ride on the world-famous Alaskan Railroad, whitewater rafting and quiet, intimate hours surrounded by sparkling glaciers, milky-blue fjords, emerald forests and majestic mountains. Nights are spent in cozy, heated wilderness lodges and cabins complete with fresh linens and hot showers. Beginning in Kenai and ending in Anchorage, participants on this trip enjoy the most beautiful of Alaska's wilderness regions, from the majestic tidewater glaciers of the Kenai Fjords National Park to the upland tundra and stunning high-mountain vistas of Denali National Park. Experienced naturalist guides lead the way, providing information about local ecology, wildlife and conservation ethics. Camping in spacious safari tents is available for the more rugged traveler.

ABOUT THE TRIP

LENGTH OF TRIP: 7, 8, 10 or 12 days

DEPARTURE DATES: 75 departures, June, July and Aug.

TOTAL COST: $1,995-$2,795 depending on length

TERMS: Credit card, checks; $50 discount if reservations made before Feb. 1, up to Mar. 1 send $400 deposit, after Mar. 1, 50% of trip fee required for deposit; cancellation fee up to 60 days prior to departure

DEPARTURE: Arrive at airport in Kenai; further transport provided by Alaska Wildland Adventures in airporter-style buses

AGE/FITNESS REQUIREMENTS: Must be at least 12 years of age; average physical fitness

SPECIAL SKILLS NEEDED: A sense of adventure and spirit of enthusiasm

TYPE OF WEATHER: Rainy and sunny days, 60°-75° in the daytime, 40°-50° at night

OPTIMAL CLOTHING: Two-piece quality raingear, lightweight and water-repellent hiking boots, wool or pile pants, shirts (one wool), warm jacket and sweater, polypropylene long underwear, wool socks and wool hat

GEAR PROVIDED: All equipment

GEAR REQUIRED: None

CONCESSIONS: Film can be purchased during trip, but purchasing in advance is recommended.

WHAT NOT TO BRING: Suitcases for packing (duffel bags preferred), poor-quality raingear

YOU WILL ENCOUNTER: An abundance of wildlife including bald eagles, moose, brown bears, caribou, Dall sheep, puffins, harbor seals, sea lions, up to three species of whales and much more

SPECIAL OPPORTUNITIES: Spectacular photography of mountains, rivers, glaciers, wildlife, marine life

ECO-INTERPRETATION: Trained naturalists interpret both cultural and natural history of Alaska throughout the trip. An emphasis is placed on wildlife, flora, geology, ecosystems and ecological concepts.

RESTRICTIONS: None

REPRESENTATIVE MENU: *Breakfast*—eggs, pancakes, French toast, cereal, fruit, juice, coffee and tea; *Lunch*—deli sandwiches and fruit; *Dinner*—fish, poultry, pasta, meats, vegetables, dessert, coffee, tea, beer and wine; *Alcohol*—complimentary wine or beer is served on occasion. Guests can purchase other liquor locally if desired.

ABOUT THE COMPANY

ALASKA WILDLAND ADVENTURES
P.O. Box 389, Girdwood, AK 99587, (800) 334-8730

For 16 years, Alaska Wildland Adventures (AWA) has offered environmentally sensitive natural-history safaris, wilderness expeditions and sportfishing packages. AWA, owned by Kirk Moessle and Jim Wells, donates 10 percent of its earnings to various environmental groups, recycles and uses environmentally sound products.

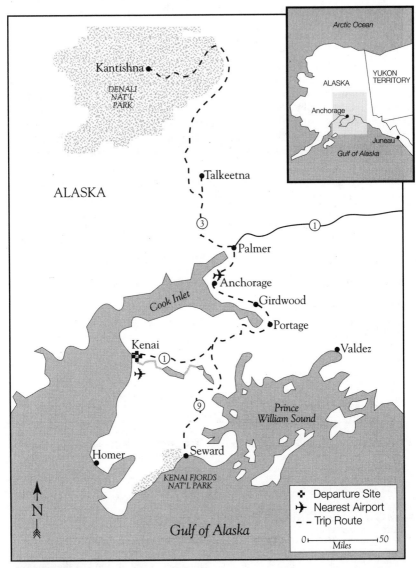

ECOFOCUS Encroachment of industry and tourism into previously wild areas is beginning to impact the land and its wildlife. Areas that had been traversed mostly by small groups practicing low-impact travel are now being overrun by large tour buses.

AROUND KOOTZANAHOO ISLAND
Explore the heart of southeast Alaska with Biological Journeys

Temperate rainforests, native cultures, snowcapped mountains, bald eagles, outpost communities, waterfalls, whales and much more await your discovery on this expedition. Known by natives as "The Fortress of the Bears," Kootzanahoo Island is estimated to be home to one grizzly bear per square mile. In addition to Kootzanahoo Island, this exploration of southeast Alaska will take you to Glacier Bay and Chichagof and Baranof islands as well as the imposing mainland. The trip begins with a visit to Icy Straits, a favorite feeding ground for minke and humpback whales. Then you'll travel to the tiny village of Tenakee Springs, which boasts magnificent gardens, rural culture and a hot bath owned and operated by local women. The native settlements of Angoon and Hoonah will expose you to some of the conflicts between native people and the US government. At Pack Creek Bear Reserve, you will enjoy the rare opportunity to observe and photograph bears up close as they fish for salmon. The waters of Frederick Sound will reveal to you the breaching, tail slapping, flipper slapping and "fluking" of over 300 humpback whales. This fantastic adventure will conclude with an unforgettable hour-long flight over nearby glaciers, mountains, snowfields and waterways surrounding Kootzanahoo Island.

ABOUT THE TRIP

LENGTH OF TRIP: 8-10 days
DEPARTURE DATES: July and Aug.
TOTAL COST: $2,995-$3,195
TERMS: Cash, check; $500 deposit; full refund minus $100 if cancellation 60 days prior to departure.
DEPARTURE: Arrival at airport in Juneau; further transport provided
AGE/FITNESS REQUIREMENTS: Average health
SPECIAL SKILLS NEEDED: A love of nature and an open mind
TYPE OF WEATHER: Chance of showers; daytime temperatures 50°-75°
OPTIMAL CLOTHING: Casual camping clothes, tennis shoes
GEAR PROVIDED: Bedding, towels, washcloths

GEAR REQUIRED: Binoculars, sunglasses, sun hat, duffel bag, raingear and knee-high rubber boots
CONCESSIONS: Stores will be available on the first and last days of the trip, but guests should come prepared.
WHAT NOT TO BRING: Hard suitcases, formal attire
YOU WILL ENCOUNTER: Some outpost communities; plenty of wildlife, including grizzly bears, humpback whales and bald eagles
SPECIAL OPPORTUNITIES: Fishing
ECO-INTERPRETATION: A world-class naturalist provides ongoing natural-history interpretation augmented by evening lectures and slide shows.
RESTRICTIONS: Minor
MEALS: California cuisine prepared by gourmet chef

ABOUT THE COMPANY

BIOLOGICAL JOURNEYS
1696 Ocean Dr., McKinleyville, CA 95521, (707) 839-0178, (800) 548-7555

For 12 years, Biological Journeys has been leading adventure-travel trips specializing in natural history and whale watching. A for-profit organization owned by Ron LeValley and Ron Storro-Patterson, the company runs expeditions that consist of small groups that tread lightly on the land while providing excellent educational opportunities for participants. The owners hope this will encourage donations and conservation activism. Biological Journeys contributes to local research activities and provides passengers with opportunities to donate directly to local research and conservation organizations. They also help environmentally concerned organizations raise funds through their company's travel programs.

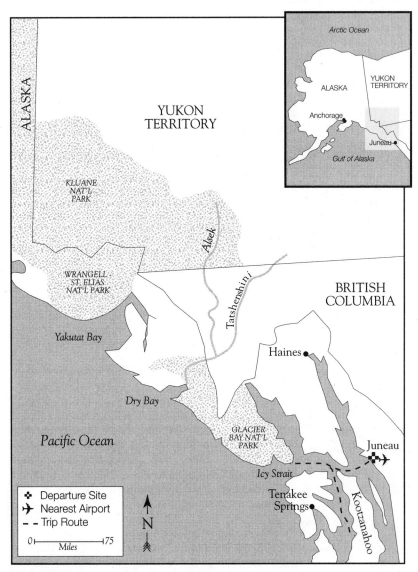

Alaska

ALASKA

YUKON
TERRITORY

KLUANE
NAT'L
PARK

Alsek

WRANGELL
ST. ELIAS
NAT'L PARK

Yakutat Bay

Tatshenshini

BRITISH
COLUMBIA

Haines

Dry Bay

GLACIER
BAY NAT'L
PARK

Pacific Ocean

Juneau

Icy Strait

Tenakee
Springs

Kootzanahoo

Inset map:
Arctic Ocean

ALASKA

YUKON
TERRITORY

Anchorage

Juneau

Gulf of Alaska

Legend:
❖ Departure Site
✈ Nearest Airport
- - Trip Route

0 |——————| 75
Miles

N

ECOFOCUS The Kootzanahoo Island region suffers from habitat destruction, primarily due to timbering. Bear hunting, development and unmanaged fishing also greatly affect the flora and fauna of the area.

THE BROOKS RANGE ADVENTURE
Backpack and raft in Alaska with Nichols Expeditions

This magical trip takes you through Alaska's Brooks Range in the heart of Gates of the Arctic National Park for an exciting combination backpack and rafting adventure. The trip begins with a charter flight from Fairbanks to Bettles and a floatplane trip from Bettles to Summit Lake, where trip participants will embark on their 15-day journey. Following the North Fork from its headwaters, the backpacking portion begins with five days of exploring pristine alpine valleys punctuated by cascading waterfalls and wildlife sightings. On the fifth day, the expedition will come to a small runway where all the river rafts and supplies will be waiting for the next leg of the trip. The North Fork is fed by numerous side-streams as it passes the majestic mountain peaks, Mt. Boreal and Frigid Crags, which Robert Marshall termed "The Gates of the Arctic." Day hikes to some of the splendid lakes and mountains in the southern part of the range round out the excursion. On the final day, the trip joins the Main Fork of the Koyukuk and continues on to Bettles for the farewell dinner.

ABOUT THE TRIP

LENGTH OF TRIP: 14 days

DEPARTURE DATES: June 21-July 4

TOTAL COST: $1,995

TERMS: Cash or VISA; $200 deposit required with full payment due 60 days prior to departure; full refund less $50 if cancelled 60 days prior

DEPARTURE: The nearest airport is Fairbanks; taxi provides transportation to Fairbanks departure point

AGE/FITNESS REQUIREMENTS: 13 years minimum; good physical condition

SPECIAL SKILLS NEEDED: Backpack experience not required, but helpful

TYPE OF WEATHER: Moderate with a possibility of rain; 40°-80° expected

OPTIMAL CLOTHING: Rainsuit, sweaters, pants, shirts, brimmed wool hat and lightweight hiking boots

GEAR PROVIDED: Rafts, first aid, cooking utensils, food and all community gear

GEAR REQUIRED: Backpack, sleeping bag, tent and thigh-high fishing waders

CONCESSIONS: Film and personal items should be purchased prior to trip.

WHAT NOT TO BRING: Radios, weapons

YOU WILL ENCOUNTER: Excellent mountain views and diverse wildlife

SPECIAL OPPORTUNITIES: Hiking, fishing and photography

ECO-INTERPRETATION: Chuck and Judy Nichols, who've guided this trip for 11 years, teach guests about the ecology and history of the area.

RESTRICTIONS: Low-impact hiking practices

REPRESENTATIVE MENU: *Breakfast*—whole-grain banana pancakes; *Lunch*—Mexican salad with tortillas; *Dinner*—fresh-grilled grayling, sourdough biscuits, carrot salad and apple cobbler; *Alcohol*—small amounts available

ABOUT THE COMPANY

NICHOLS EXPEDITIONS
497 N. Main St., Moab, UT 84532, (800) 635-1792, (801) 259-7882
Active in educating guests about regional environmental issues, Nichols Expeditions also contributes to conservation groups, such as the Utah Wilderness Alliance, and writes letters asking for protection of rivers and wilderness areas, such as the Copper River in Alaska. This for-profit outfitter, owned by Chuck and Judy Nichols since 1979, specializes in outdoor adventure travel.

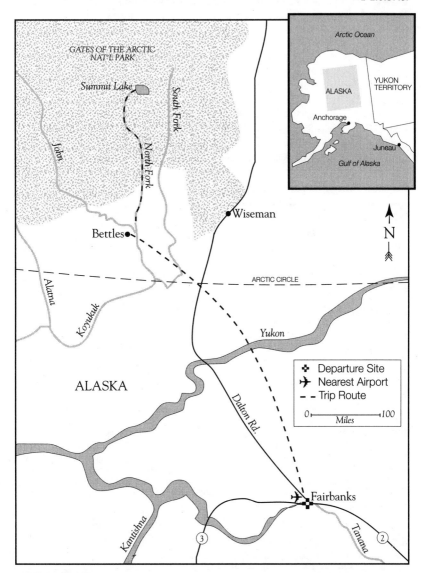

ECOFOCUS The prime environmental threat to the region stems from the proposed oil and gas exploration in the adjacent Arctic National Wildlife Refuge and the monumental development that would accompany it. An operation of this scale would produce roads and pipelines and have significant impact on the nearby Brooks Range and its wildlife.

RAFTING THE TATSHENSHINI-ALSEK RIVER
Raft and hike with James Henry River Journeys/Wilderness Journeys

Known as North America's most pristine and magnificent river, the Tatshenshini-Alsek River in Alaska is the setting for an unforgettable and picturesque 14-day river trip. This excursion will take trip participants through the exquisitely beautiful St. Elias and Fairweather Ranges, site of the world's largest nonpolar glacial system and several 15,000-foot peaks. The trip begins on a calm, meandering stream that broadens to a lively, Class III-IV whitewater river as it is fed by hundreds of tributaries. Scores of multi-hued wildflowers, mosses and berries line the banks, while fir, spruce and hemlock trees make deep-green forests. Wildlife abounds in this remote habitat. Bald eagles soar overhead as grizzly and black bears roam the shoreline. Alaskan moose, mountain goats, Dall sheep, wolves, gray foxes, flocks of spotted sandpipers and Canada geese can also be seen. Four layover days during the trip allow adventurers time to explore scenic mountain valleys and hike the high alpine tundra and glaciers near camp. At night, participants gather around the campfire, read Alaskan wilderness literature and Tlingit oral narratives and discuss brown bear mythology.

ABOUT THE TRIP

LENGTH OF TRIP: 14 days

DEPARTURE DATES: July-Aug.

TOTAL COST: $2,170

TERMS: Check, VISA, MC; call outfitter for details about reservations and cancellations

DEPARTURE: Arrive at Juneau airport; travel by air or ferry to Haines departure point.

AGE/FITNESS REQUIREMENTS: 14 years and older

SPECIAL SKILLS NEEDED: None

TYPE OF WEATHER: 60°-70° during the day, possibility of rain

OPTIMAL CLOTHING: Wool and/or fleece garments, waterproof raingear and above-the-calf rubber boots

GEAR PROVIDED: Tents, life jackets, waterproof bags for personal gear and ammunition cans for cameras

GEAR REQUIRED: Sleeping bags, pads, raingear, rubber boots and day packs

CONCESSIONS: Film, souvenirs and personal items are available in Haines.

WHAT NOT TO BRING: Firearms, radios

YOU WILL ENCOUNTER: Wildlife

SPECIAL OPPORTUNITIES: Hiking onto ridges above the river and walking on glacial moraines

ECO-INTERPRETATION: During the trip, naturalists will share with trip participants their insights into geology, botany and wildlife of this dynamic river valley.

RESTRICTIONS: Low-impact camping and consideration for historical artifacts

REPRESENTATIVE MENU: *Breakfast*—eggs, French toast, multi-grain cereals, fruit and yogurt; *Lunch*—salads, fruit, cheese, cold cuts, nuts and juices; *Dinner*—lasagna, barbecued lamb, clam linguine and Irish stews; *Alcohol*—bring your own, if desired

ABOUT THE COMPANY

JAMES HENRY RIVER JOURNEYS/WILDERNESS JOURNEYS
P.O. Box 807, Bolinas, CA 94924, (800) 786-1830
James Henry River Journeys/Wilderness Journeys, a 20-year-old, for-profit outfitter owned by Jim Katz, is a concessionaire of Glacier Bay National Park and Preserve specializing in adventure trips in North America and the Far East. Katz has authored articles for various environmental magazines, specifically to stop the proposed mining near the Tatshenshini River. He encourages trip guests to write letters relating to local water policy issues.

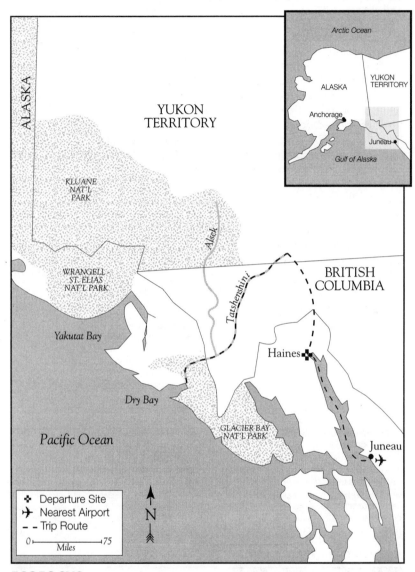

ECOFOCUS A Canadian mining company is currently threatening to mine 100 million tons of ore containing copper, gold, silver and cobalt from Windy Craggy Mountain near the confluence of the Tatshenshini and Alsek rivers. Any leaching of ore, which has a high sulfide content, into the river would contaminate the watershed and destroy the large salmon runs that occur every year. Large grizzlies, Dall sheep and the Chilkat Bald Eagle Preserve would also be disturbed by the project.

SAILING THE SOUND
Sail Prince William Sound with Alaskan Wilderness Sailing Safaris

While Prince William Sound is sometimes thought of as merely a transportation route for large oil tankers, the sheer beauty of its snow-clad peaks and glaciers, which descend sharply down to sea level, will surely leave a more pleasing impression. This sailing trip takes you through the heart of the sound, which, in spite of the oil spill, still retains complete habitats for its dazzling array of wildlife. Marine mammal watching opportunities abound on this trip, as participants discreetly watch Stellar sea lions sleeping or socializing, harbor seals carried out to sea on icebergs, and sea otter mothers parenting their pups. Experience the thrill of spotting humpback, killer and minke whales. All trips include a visit to the Columbia Glacier, the last glacier in North America to undergo a drastic retreat following the end of the neo-glacial period. Here, guests will be able to walk among icebergs and discover remnants of old forests frozen in the glacier, as well as stroll across land only recently uncovered by the glacier. Sea kayaks are also available for rental to explore the shorelines and shallow nooks and crannies of the sound. Solitary or guided walks along the sound's rich intertidal zones or into the ancient old growth forests help round out the trip and introduce guests to another unique ecosystem.

ABOUT THE TRIP

LENGTH OF TRIP: 3-9 days

DEPARTURE DATES: May 15-Sept. 15

TOTAL COST: $200/person/day

TERMS: Checks only; $300 deposit to reserve; full refund less 15% of total trip cost if cancelled 30 days or more prior to trip departure

DEPARTURE: Nearest airport in Valdez, AK; shuttle by tour boat from airport to the Growler Island departure point

AGE/FITNESS REQUIREMENTS: None

SPECIAL SKILLS NEEDED: None

TYPE OF WEATHER: Variable weather with temperatures 50°-70°

OPTIMAL CLOTHING: Raingear and pile clothing

GEAR PROVIDED: Kayaks, raingear and sleeping bag

GEAR REQUIRED: All personal gear

CONCESSIONS: All personal items can be purchased in Valdez.

WHAT NOT TO BRING: Firearms

YOU WILL ENCOUNTER: Diverse wildlife, including bears, eagles, sea otters, seals, Stellar sea lions, porpoises and whales; tidewater glaciers and ancient coastal rainforest

SPECIAL OPPORTUNITIES: Photography, solitary or guided walks and sea kayaking

ECO-INTERPRETATION: Guides provide informative talks on recovery from the 1989 oil spill, effects of 1964 earthquake, marine mammals, birds, glaciers, geology and plant communities.

RESTRICTIONS: None

REPRESENTATIVE MENU: *Breakfast*—pancakes, fruit, cereal, yogurt, toast and eggs; *Lunch*—homemade soup, sandwiches, salads and dessert; *Dinner*—chicken or fish entrees, salad, rice/potatoes and/or rolls and dessert; *Alcohol*—not provided; guests may bring for use while boats are anchored

ABOUT THE COMPANY
ALASKAN WILDERNESS SAILING SAFARIS
Box 1313 BW, Valdez, AK 99686, (907) 835-5175

Owned by Nancy and Jim Lethcoe, Alaskan Wilderness Sailing Safaris (AWSS) has been in business for 18 years with a staff of two, specializing in sailing and sea kayaking trips. The Lethcoes have worked on issues affecting the sound's forests, marine mammals and land management for over 20 years. During the *Exxon Valdez* oil spill, Nancy represented environmental interests on the InterAgency Shoreline Cleanup Committee and served as executive director of the Prince William Sound Conservation Alliance. AWSS was the only for-profit company authorized to do volunteer cleanup in Prince William Sound. The Lethcoes have written seven books on the natural history of Prince William Sound.

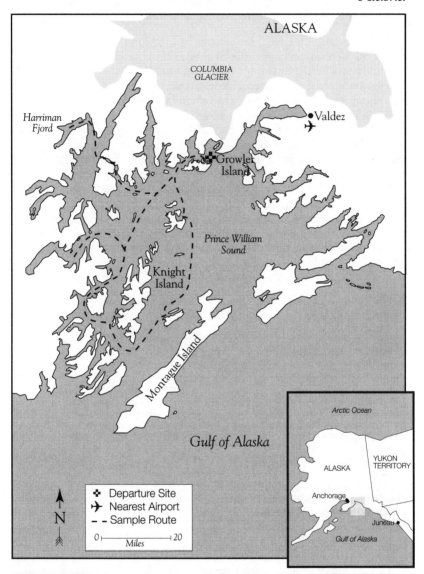

ECOFOCUS Prince William Sound is still recovering from the 1989 *Exxon Valdez* oil spill. Although oil companies have made significant changes to improve the safety of oil transport, the risks from fire and adverse winter weather conditions still remain. Potential activities that threaten the sound's future include developments in the tourism, fisheries, timber and mining industries.

ARIZONA DESERT WILDLIFE RESEARCH
Study Gila Valley wildlife with the Foundation for Field Research

The Foundation for Field Research is looking for a few good volunteers. Come join researchers at the Sears Point archaeological site in southwest Arizona as they attempt to document the abundance of various species, both plant and animal, to determine the fluctuations of the area's native population. What is thought of as a land of desolation and stillness is, in reality, an area that supports a very diverse and active wildlife population. At Sears Point, a forest of mesquite and palo verde provides a home for numerous birds and mammals such as bobcats, cottontail rabbits, deer, woodpeckers, wrens, doves and ravens, while chipmunks, ringtail cats, wild honeybees, badgers and various lizards find refuge in the nearby majestic volcanic mesas. Volunteers with binoculars, cameras and patience will set up and check a trap line of Sherman and Havahart live-animal traps, so the animals can be measured, photographed, inspected and weighed. Mist nets will be used to gently trap birds and observation blinds will be placed on the mesa top to view the flats below for deer. None of the traps harm the animals; they will be released soon after they are caught. A working knowledge of birds, mammals and small reptiles is helpful, but not necessary.

ABOUT THE TRIP

LENGTH OF TRIP: 7 days
DEPARTURE DATE: Dec. 21
TOTAL COST: $495
TERMS: Pay by check; $250 deposit
DEPARTURE: Trip starts in Phoenix, AZ
AGE/FITNESS REQUIREMENTS: Minimum 14 years old
SPECIAL SKILLS NEEDED: None
TYPE OF WEATHER: Warm, possible rain
OPTIMAL CLOTHING: Warm clothing, boots
GEAR PROVIDED: Tent, sleeping pad
GEAR REQUIRED: Sleeping bag
CONCESSIONS: Film and personal items available in Phoenix

WHAT NOT TO BRING: Electrical appliances
YOU WILL ENCOUNTER: Coyotes, mountain lions, ringtail cats, other mammals and many species of birds
SPECIAL OPPORTUNITIES: Excellent photography opportunities
ECO-INTERPRETATION: By helping with research, participants learn much about local wildlife.
RESTRICTIONS: Must not touch relics when camping at prehistoric religious site
REPRESENTATIVE MENU: *Breakfast*—coffee, bacon, pancakes; *Lunch*—sandwiches; *Dinner*—Dutch-oven cooking—chicken, beef, desserts; *Alcohol*—available

ABOUT THE COMPANY

FOUNDATION FOR FIELD RESEARCH
P.O. Box 2010, Alpine, CA 91903, (619) 445-9264
The Foundation for Field Research, a 10-year-old nonprofit company, runs expeditions in Europe, Mexico, the Caribbean, West Africa and the western US to log vital environmental data. With a staff of nine, the foundation looks for volunteers willing to contribute money and time to aid important environmental research.

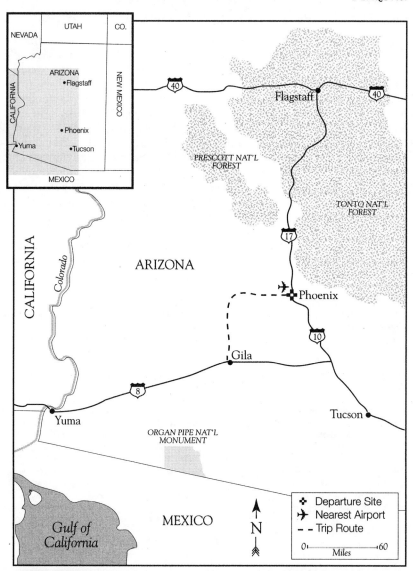

ECOFOCUS The ecosystem of the Gila River has been greatly altered due to several dams that divert much of its water for agricultural purposes. The lack of water has significantly reduced the numbers of some animals, such as deer and mountain lions, as well as nearby bird and plant life.

GRAND CANYON RAFTING TRIP
Float the magnificent Grand Canyon with Arizona Raft Adventures

Arizona Raft Adventures offers a wide variety of trips through the Grand Canyon on oar, paddle and motorized rafts. Regardless of the type of vessel you choose, rafting down this extraordinary stretch of the Colorado River is undoubtedly the most exciting and thorough way to see the canyon. Your day starts early with a hearty breakfast prepared by guides. You will spend the day crashing through roaring rapids or lazily drifting down still water, dozing in the sun. Your Grand Canyon adventure is yours to create. Try your hand at rowing or paddling, learn about the rich natural history, help in the preparation of meals, explore the many side canyons or sit alone by the river. As the sun sinks behind the cliffs, pull over onto a sandy beach and set up camp. Dinners are fresh and delicious and the stars overhead wink you to sleep. Experience not only the river but the secrets of the canyon, as revealed by expert guides, some of whom have spent decades on the river. Without a doubt, this experience should not be missed. Whether your vacation lasts six days or two weeks, this raft trip will expose you to one of eight natural wonders of the world and teach you what you can do to preserve it.

ABOUT THE TRIP

LENGTH OF TRIP: 6-14 days

DEPARTURE DATES: Approximately 50 during the spring, summer and fall

TOTAL COST: $986 - $1,884

TERMS: Cash, check; $200 deposit within 10 days of booking; cancellation before 90 days prior to trip

DEPARTURE: Arrival at airport in Flagstaff; transport to and from river provided

AGE/FITNESS REQUIREMENTS: Must be at least 10 years old

SPECIAL SKILLS NEEDED: None

TYPE OF WEATHER: Warm desert climate, daytime temperatures 90°-100°

OPTIMAL CLOTHING: Lightweight cotton clothes for days, warm clothes for night, raingear, swimsuit, sneakers or river sandals, hat

GEAR PROVIDED: Rafting gear and waterproof bags, sleeping units on partial trips

GEAR REQUIRED: Tent and sleeping bag (both can be rented)

CONCESSIONS: A retail outlet is available for last-minute needs.

WHAT NOT TO BRING: Weapons, lawn chairs, or valuables

YOU WILL ENCOUNTER: Exceptional wildlife including desert bighorn sheep, great blue herons and ibises

SPECIAL OPPORTUNITIES: Hiking through side canyons, learning to row and paddle

ECO-INTERPRETATION: All guides are knowledgeable about the natural history and geology of and environmental threats to the Grand Canyon.

RESTRICTIONS: All fecal matter is packed out.

REPRESENTATIVE MENU: *Breakfast*—eggs, bacon, yogurt, toast, cereal; *Lunch*—sandwiches, fruit, cheese, meat, cookies; *Dinner*—steak, potatoes, vegetables, chicken, pasta, soup, cake; *Alcohol*—bring your own if desired

ABOUT THE COMPANY

ARIZONA RAFT ADVENTURES
4050-T E. Huntington Dr., Flagstaff, AZ 86004, (800) 786-7238
Specializing in adventure travel and whitewater rafting, Arizona Raft Adventures (AzRA) is an environmentally conscious company in business for 18 years. Some 10 percent of its pre-tax profits are donated to the Grand Canyon Trust, American Rivers and various local causes. President of AzRA Rob Elliot devotes much of his time to the protection of the Grand Canyon and serves as a spokesperson for the river recreation community.

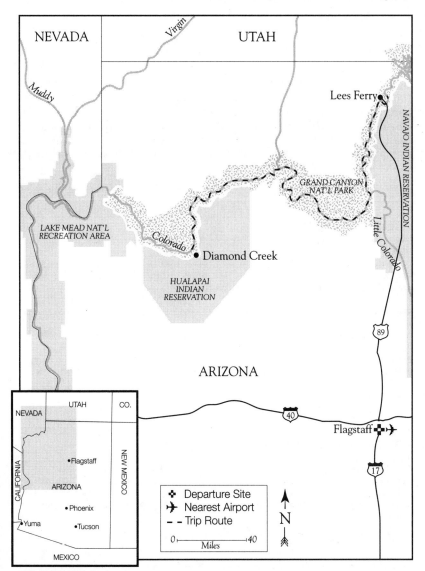

ECOFOCUS The Glen Canyon Dam, which holds back Lake Powell and lies just upstream from the Grand Canyon, discharges greatly fluctuating amounts of water that cause beach erosion during high flows. The increased water flow, released to meet peak electricity demands, also endangers a multitude of fish and wildlife. Another environmental problem in the area is the pollution generated by the nearby Navajo Power Plant, which creates hazy conditions around the canyon.

ANACAPA ISLAND TRIP
Camp on the Channel Islands with Island Packers

The emphasis of the Anacapa Island Trip is "packing in" small groups of people to the islands so as not to damage the island habitat. It usually takes an hour and a half to cross the channel to the Anacapa Islands. Then, after an exciting island landing that is occasionally wet, guests pass into the domain of the tiny islands rife with opportunities for recreation and education. The island offers some hiking and panoramic views. Whether you're looking for solitude or group interaction, you've come to the right place. A National Park Service ranger is usually available to answer any questions and lead nature walks. Beautiful clear waters and kelp beds teeming with sea life provide hours of swimming and snorkeling. Bring the family for a relaxing day on a beautiful island full of rich wildlife away from crowded mainland beaches.

ABOUT THE TRIP

LENGTH OF TRIP: 2-4 days

DEPARTURE DATES: Daily June-Sept.

TOTAL COST: $48/adult, $30/child

TERMS: All forms accepted; full payment within 7 days of phone confirmation; full refund if cancelled 5 days or more prior to departure

DEPARTURE: Arrive at Los Angeles or Oxnard; shuttle to Ventura Harbor departure point

AGE/FITNESS REQUIREMENTS: Good physical condition

SPECIAL SKILLS NEEDED: None

TYPE OF WEATHER: Sunshine and breezy afternoons, ranging from 70°-85°

OPTIMAL CLOTHING: Layers of old clothes, jacket, light hiking boots or tennis shoes

GEAR PROVIDED: None

GEAR REQUIRED: Water, food, tent, stove and snorkeling gear

CONCESSIONS: Film can be purchased on the trip

WHAT NOT TO BRING: Heavy items (no more than two scuba tanks)

YOU WILL ENCOUNTER: Extensive marine life including marine birds, California sea lions, harbor seals, dolphins, etc.

SPECIAL OPPORTUNITIES: Snorkeling and scuba diving

ECO-INTERPRETATION: Naturalists provide information throughout the day and there are a ranger and a naturalist on the island.

RESTRICTIONS: None

MEALS: Meals are not provided.

ABOUT THE COMPANY

ISLAND PACKERS
1867 Spinnaker Dr., Ventura Harbor, CA 93001, (805) 642-1393

Running trips to the Channel Islands for 24 years, Island Packers is a for-profit outfitter specializing in recreation and research-oriented travel. They educate guests about preservation and donate funds to the Nature Conservancy and National Park Service to help protect the islands and marine life.

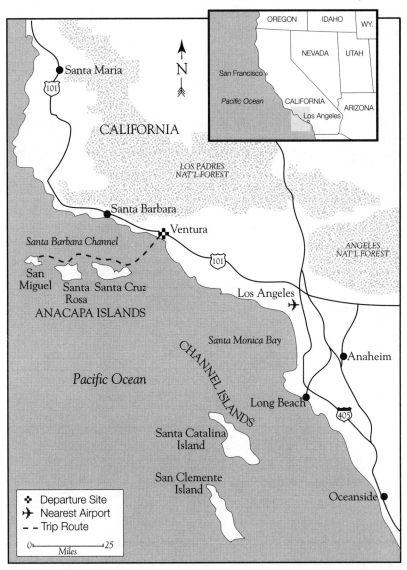

California

OREGON *IDAHO* *WY.*

NEVADA *UTAH*

San Francisco

Pacific Ocean CALIFORNIA *ARIZONA*

Los Angeles

Santa Maria

N

CALIFORNIA

101

LOS PADRES NAT'L FOREST

Santa Barbara

Ventura

Santa Barbara Channel

101

San Miguel Santa Rosa Santa Cruz

ANGELES NAT'L FOREST

ANACAPA ISLANDS

Los Angeles

Santa Monica Bay

CHANNEL ISLANDS

Pacific Ocean

Anaheim

Long Beach

405

Santa Catalina Island

San Clemente Island

Oceanside

✤ Departure Site
✈ Nearest Airport
- - Trip Route

0 ⊢——————⊣ 25
Miles

ECOFOCUS The Channel Islands are well protected by guidelines that regulate the number of visitors allowed to the islands. Nearby oil platforms pose the danger of oil spills, which would devastate the animal life on and around the islands.

THE TUOLUMNE RIVER
Raft the Tuolumne River with ECHO: The Wilderness Company

Flowing from the high reaches of Yosemite National Park, the Tuolumne River is quickly slowed by the Hetch-Hetchy Reservoir at the edge of Tuolumne Park and then released downstream, forming one of the most exciting and whitewater-filled stretches of river in the US. Set amid the grand and unspoiled Sierra Canyon, the Tuolumne thrills guests with thundering rapids such as Sunderland's Chute and the legendary, world-class Clavey Falls as it drops 66 feet per mile for the first six miles and 38 feet per mile overall. The river takes on many different faces, depending on the time of year you choose to go. In May and June, during peak snowmelt, the river is high and wild. When the river recedes later in the year, skillful paddling is required to maneuver around newly exposed obstacles. In addition to the fantastic whitewater, the Tuolumne canyon offers plenty of breathtaking and rugged wilderness. Hikes up side creeks reveal beautiful waterfalls and calm swimming holes that are refreshing refuges during the hot summer months. For the adventurous, more strenuous hikes up the main canyon reward hikers with dramatic views of the Tuolumne and also take them to old mining sites. The crystal-clear waters of the Tuolumne also offer some of the finest trout fishing around.

ABOUT THE TRIP

LENGTH OF TRIP: 1-3 days

DEPARTURE DATES: Mar.-Sept.

TOTAL COST: $150-$398

TERMS: Check and credit cards accepted; deposit of $100 (or half of total fees, whichever is less); $50 cancellation fee if made prior to final payment date

DEPARTURE: Arrive in either Sacramento, Oakland or San Francisco; car rental available to Groveland, CA, departure point

AGE/FITNESS REQUIREMENTS: 12 years minimum; good health

SPECIAL SKILLS NEEDED: None

TYPE OF WEATHER: Spring can be cold and rainy; summer can be hot in the canyons with temperatures in the 90s and 100s.

OPTIMAL CLOTHING: Swimsuit or shorts, tennis shoes or sandals (Tevas), raingear and polypropylene underwear

GEAR PROVIDED: Boats, paddles, life jackets, waterproof bags

GEAR REQUIRED: Clothing, sleeping bag and tent

CONCESSIONS: Concessions can be purchased in Groveland, but not on the trip.

WHAT NOT TO BRING: Radios and firearms

YOU WILL ENCOUNTER: Extensive wildlife including eagles, hawks, deer and trout

SPECIAL OPPORTUNITIES: Fishing, photography and hiking

ECO-INTERPRETATION: All guides are knowledgeable about the environmental, natural and social history of the area.

RESTRICTIONS: None

REPRESENTATIVE MENU: *Breakfast*—cereal, yogurt, fresh fruit, scrambled eggs, bacon, muffins, juice and coffee; *Lunch*—deli sandwiches, cheeses, vegetables, fresh fruits and cookies; *Dinner*—barbecued marinated chicken, rice, vegetable stir-fry, fresh salad, Dutch-oven-prepared dessert; *Alcohol*—not provided, but can be bought

ABOUT THE COMPANY

ECHO: THE WILDERNESS COMPANY
6529 Telegraph Ave., Oakland , CA 94609, (510) 652-1600

Offering low-impact river camping trips in a remote part of the historic Sierra California foothills for the last 20 years, ECHO: The Wilderness Company strives to promote a greater appreciation of wilderness and wild rivers through education. The for-profit company, owned by Joe Daly and Dick Linford, also donates $1 per passenger per day to the Tuolumne River Preservation Trust and Friends of the River.

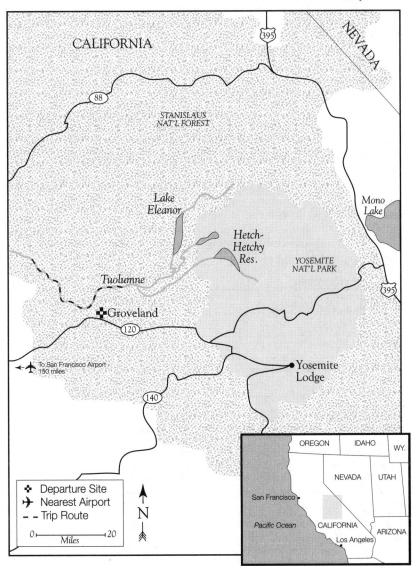

ECOFOCUS The Clavey River, the largest free-flowing tributary of the Tuolumne, is being threatened with the construction of five proposed dams. Joe Daly, co-owner of ECHO and president of the Tuolumne River Preservation Trust, has volunteered many hours toward stopping this and other damaging projects, as well as working to get the Clavey Wild and Scenic River designation.

DOLORES RIVER
Raft the Dolores River with Bill Dvorak Kayak and Rafting Expeditions

What the Dolores River lacks in notoriety it makes up for in splendor. The river, which takes a backseat only to the Colorado River through the Grand Canyon and the Middle Fork of the Salmon River in unbroken miles of river running, is in an area not heavily touched by human hands and offers miles of terrain that remains virtually unexplored. Beginning high in the San Juan Mountains at about 8,000 feet, this unique river run treats you to towering forests of ponderosa pines and Douglas firs and thrills you with its famous Snaggletooth Rapid. Next, the river winds through Slickrock Canyon, where redtail hawks soar, then slips into Paradox Valley, where cottonwoods provide shade and refuge for nesting birds. Along the way, you'll camp under overhangs used by Dolores inhabitants a thousand years ago and view prehistoric pictographs and Indian sites. The final leg of the Dolores before it meets the Colorado River is dotted with rapids created by flash floods from side canyons. Then it is an easy float to the historic take-out point at Dewey Bridge in the Castle Valley, for the conclusion of this memorable trip.

ABOUT THE TRIP

LENGTH OF TRIP: 3-12 days

DEPARTURE DATES: Late Apr.-mid-June

TOTAL COST: $310-$1,250

TERMS: VISA, MC, AmEx, check, cash; $100-$450 deposit to reserve; full refund less $50 if cancelled 30 days or more prior to trip

DEPARTURE: Can be picked up at airport in Cortez, CO, or meet at put-in near Cahone, CO

AGE/FITNESS REQUIREMENTS: Minimum of 10 years; reasonable health

SPECIAL SKILLS NEEDED: None

TYPE OF WEATHER: Cool (55°-70°) in the spring and hot (80°-90°) in the summer

OPTIMAL CLOTHING: Casual clothing, shorts, tennis shoes, sandals

GEAR PROVIDED: All river and safety equipment, cooking and eating utensils, raingear; rental camping gear available

GEAR REQUIRED: Personal clothing, camping equipment, individual beverage requirements

CONCESSIONS: Film, personal items and souvenirs can be purchased in Cortez or at the company shop, but not during the trip.

WHAT NOT TO BRING: Electrical appliances, radios and weapons

YOU WILL ENCOUNTER: Various wildlife, including deer, elks, bears, beavers, big horn sheep, eagles, hawks, herons and geese

SPECIAL OPPORTUNITIES: Kayaking instruction, inflatable kayaks, visits to ancient Anasazi ruins and special music performances

ECO-INTERPRETATION: Guides provide interpretive information throughout the trip on flora, fauna, geology, natural history and environmental problems.

RESTRICTIONS: Observation of Antiquities Act at all natural historic sites

REPRESENTATIVE MENU: *Breakfast*—fresh fruits, cereal, yogurt, various egg dishes, bacon, toasted muffins, fruit juice, coffee and tea; *Lunch*—buffet-style deli sandwiches with cold meats, cheeses, salad, fresh fruits, juices, soda, cookies and chips; *Dinner*—steak, baked salmon, Mexican chicken, spaghetti, sauteed vegetables, baked desserts, coffee and complimentary wine; *Alcohol*—complimentary wine with dinner; bring additional in cans or plastic containers

ABOUT THE COMPANY

BILL DVORAK KAYAK AND RAFTING EXPEDITIONS
17921 US Highway 285, Ste. EZ, Nathrop, CO 81236, (800) 824-3795
Bill and Jaci Dvorak have been outfitting since 1969. Presently, Dvorak's Expeditions specializes in running the great rivers of the western US. When the Dvoraks are not busy shooting the rapids, they support conservation organizations, such as Colorado River Outfitter Association, and work to preserve endangered American rivers.

ECOFOCUS Both the Dolores River and the Gunnison River are currently being threatened by various water interests that propose to dam or divert river water. The Dolores is being threatened by the McPhee Dam, which has put a strain on the river and its resources and has affected fishery and wildlife. Both rivers have been proposed for protected status as Wild and Scenic Rivers.

SKI MOUNTAINEERING EXPEDITION
Skiing and winter camping with Colorado Outward Bound School

Ski and camp among Colorado's snowcapped 14,000-foot peaks. The expedition covers both the Mt. Massive and Collegiate Peaks wilderness, and begins from the base camp west of the historic town of Leadville. The course, while one of the most demanding that Colorado Outward Bound School offers, is also among the most rewarding. The first few days teach skills essential for back-country survival, including avalanche awareness, winter navigation and winter camping. Then it's off to the winter wonderland of the Rocky Mountains for 12 days in the snowy and awe-inspiring mountains. Instructors will offer lessons on how to build and sleep in snow shelters, how to evaluate snow conditions and how to ski untracked powder. The daily itinerary includes rising and shining early, skiing three to five miles per day, then setting up camp in the early afternoon. Afternoon options include carving some turns in the knee-deep snow that has made Colorado famous, attempting a peak ascent or ice climbing. A 48-hour solo allows course participants an opportunity to test their new skills and experience self-reliance in a new environment away from guided instruction.

ABOUT THE TRIP

LENGTH OF TRIP: 18 days

DEPARTURE DATES: Jan.-Apr.

TOTAL COST: $1,450

TERMS: All forms of payment accepted; $60 non-refundable deposit; full refund for cancellation if more than 30 days prior to course

DEPARTURE: Arrive at Denver airport; travel by bus to Leadville departure point.

AGE/FITNESS REQUIREMENTS: At least 14 years old

SPECIAL SKILLS NEEDED: None

TYPE OF WEATHER: Cold, snow and sun

OPTIMAL CLOTHING: Complete list furnished upon registration

GEAR PROVIDED: All technical and camping gear

GEAR REQUIRED: Boots, gloves, wool pants and long underwear

CONCESSIONS: Film, souvenirs and personal items can be purchased in Leadville, but not during the expedition.

WHAT NOT TO BRING: Drugs, alcohol or cigarettes

YOU WILL ENCOUNTER: Superb wildlife viewing

SPECIAL OPPORTUNITIES: Climbing one or two major peaks; ice climbing; experiencing a 48-hour solo, weather and all conditions permitting

ECO-INTERPRETATION: Natural history, including archaeology, geology, flora and fauna, are discussed.

RESTRICTIONS: None

REPRESENTATIVE MENU: *Breakfast*—hot drinks, granola, oatmeal, eggs and muffins; *Lunch*—hot drinks, crackers and cheese, fruit and gorp; *Dinner*—hot drinks, vegetable stir-fry, spaghetti, stew and soups; *Alcohol*—none allowed

ABOUT THE COMPANY

COLORADO OUTWARD BOUND SCHOOL
945 Pennsylvania St. , Denver, CO 80203-3198, (303) 837-0880, (800) 477-2627
Colorado Outward Bound School, a nonprofit company specializing in adventure travel in the western US, has been in existence for 30 years. While providing exceptional wilderness education, the school also takes part in service projects, such as trail maintenance, in conjunction with land-management agencies.

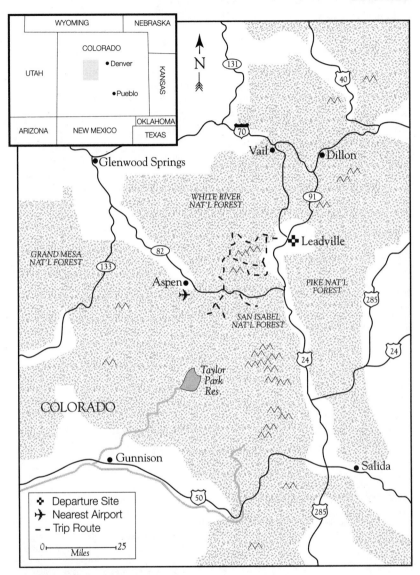

ECOFOCUS While there are no logging or mineral interests threatening the terrain covered by the ski mountaineering trip, the issue of multiple use in a national forest is an ever-present problem. Snowmobiles, for example, generate a lot of noise and disrupt the wildlife and serenity in this area.

LITTLE ST. SIMONS ISLAND
Explore the ecology of a barrier island in Georgia

Designed to be as strenuous or as relaxing as you desire, a visit to Little St. Simons Island uniquely blends undeveloped barrier island wilderness and comfortable country inn accommodations. There are no set itineraries or deadlines to meet on this 10,000-acre island. You can sit on a porch overlooking a river, lie on the beach, ride a horse, canoe, fish, boat, hike or birdwatch. Guides lead daily nature walks, during which they discuss natural history and island ecology. Touring the island by foot, boat or one of the only four vehicles on the island, you will see an incredible array of bird life, including shorebirds, wading birds, songbirds and raptors. Other wildlife on the island includes alligators and deer. After a day strolling along the sandy beaches or through the maritime forests, you will enjoy relaxing in a cozy cottage or in the lodge with one of the other travelers. Each meal is served family-style, allowing you to interact with other visitors or keep to yourself as you explore this unique coastal wilderness area.

ABOUT THE TRIP

LENGTH OF TRIP: Flexible according to group or individual requirements

DEPARTURE DATES: Individual bookings accepted Mar., Apr., May, Oct., Nov.; group bookings, June-Sept.

TOTAL COST: $300-$400 per couple per day; $2,000-$3,000 per group per day

TERMS: Checks, VISA or MC; one night deposit to reserve; full refund 30 days prior to arrival

DEPARTURE: Nearest airport is Jacksonville. Arrive at Hampton River Club Marina; ferries run twice daily to the island.

AGE/FITNESS REQUIREMENTS: None

SPECIAL SKILLS NEEDED: None

TYPE OF WEATHER: Warm with daytime temperatures 60°-90° and nights 35°-60°

OPTIMAL CLOTHING: Casual and informal clothing, footwear for muddy, wet terrain

GEAR PROVIDED: Linens and beach towels, canoes, motorboats, fishing gear

GEAR REQUIRED: Personal clothing, books and insect repellent

CONCESSIONS: T-shirts and hats are available; travelers should arrive with film and necessary personal products.

WHAT NOT TO BRING: Fancy clothes or shoes

YOU WILL ENCOUNTER: Alligators, deer and incredible bird life

SPECIAL OPPORTUNITIES: Horseback riding, hiking, crabbing, canoeing, boating and birdwatching

ECO-INTERPRETATION: Guides provide interpretive information on island history, local ecosystems and tips for birdwatching.

RESTRICTIONS: No TV or telephones

MEALS: Family-style dining

ABOUT THE COMPANY
LITTLE ST. SIMONS ISLAND
P.O. Box 1078, St. Simons, GA 31522, (912) 638-7472
In business for 12 years, Little St. Simons Island is a unique blend of undeveloped barrier island with comfortable country inn experience. Little St. Simons Island is privately owned by a family from California and managed by Debbie and Kevin McIntyre. The guides are local naturalists. The lodge proceeds are used to preserve this natural island wilderness, which is one of the last privately owned, undeveloped barrier islands off the coast of Georgia. The lodge also makes donations to National Audubon Society and to the Center for Marine Conservation.

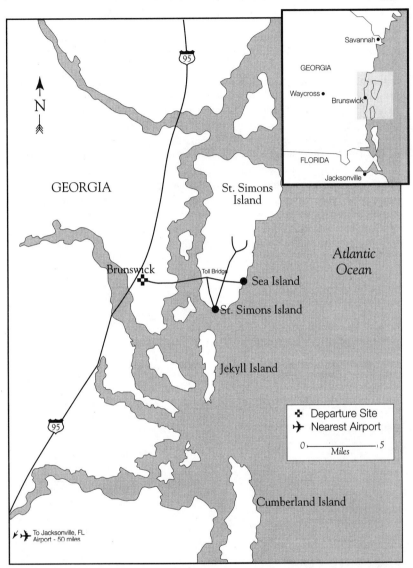

ECOFOCUS Little St. Simons Island is one of the last privately owned, undeveloped coastal islands off Georgia. Preservation of this island and of the surrounding wetlands is essential to the survival of this unique wilderness area and habitat for endangered species.

OKEFENOKEE WILDERNESS TOUR
Canoe the Okefenokee Swamp with Wilderness Southeast

Canoeing is the optimum form of transportation for fully discovering the wild and diverse beauty of the Okefenokee Wilderness. While paddling through the swamp, trip participants will come across sandy pine islands, spacious peat prairies and commanding cypress forests, as well as gain insight into the history of the early swampers by visiting a restored homestead. The pace is unhurried as guests canoe along aquatic trails shared with such swamp creatures as alligators and yellow-bellied sliders. Camping is both on the periphery of the Wilderness Area and on platforms or islands within the swamp. Sleeping on a platform in the middle of the swamp, listening to the nighttime sounds of frogs singing, alligators bellowing, sandhill cranes trumpeting and barred owls hooting, is one of the many treats of the trip. On the seven-day program, guests will paddle through an enchanting mistletoe-tupelo forest on their way to the Suwannee River, which drains the Okefenokee. Here, they will experience the subtle contrast between the swamp, with its floating peat prairies and still waters, and the river, with its gentle current, Ogeeche lime trees and white sand beaches.

ABOUT THE TRIP

LENGTH OF TRIP: 5 or 7 days

DEPARTURE DATES: Oct.-Nov., Feb.-May

TOTAL COST: $385, $530

TERMS: Checks accepted; 50% of fee required with registration; 75% refund if cancelled more than 30 days prior

DEPARTURE: Nearest airports are Jacksonville, FL and Savannah, GA; departure from Savannah airport or Suwannee Canal Recreation Area

AGE/FITNESS REQUIREMENTS: Minimum age recommended is 14; good health

SPECIAL SKILLS NEEDED: None

TYPE OF WEATHER: Warm to hot days and cool to warm nights, ranging from 30° to 70°; rain always a possibility

OPTIMAL CLOTHING: Old, comfortable clothes that can be layered

GEAR PROVIDED: Tents, canoes, food and all common gear

GEAR REQUIRED: All personal gear; sleeping bag and pad can be rented

CONCESSIONS: Bring all film and personal items with you.

WHAT NOT TO BRING: Radios, Walkmans, pets, drugs

YOU WILL ENCOUNTER: Alligators, sandhill cranes, woodstorks, raccoons, snakes, turtles, fish, ospreys

SPECIAL OPPORTUNITIES: Fishing, stargazing, listening to alligators and photography

ECO-INTERPRETATION: Explanation of natural history and environmental elements of the Okefenokee are the purpose of the trip.

RESTRICTIONS: No swimming, no dumping of trash or human wastes

REPRESENTATIVE MENU: *Breakfast*—grapefruit slices, banana pecan pancakes, sausage and coffee; *Lunch*—turkey sandwiches, nut mix, celery, carrot sticks and lemonade; *Dinner*—beef and/or bean burritos, raspberry yogurt pie and red wine; *Alcohol*—wine is provided

ABOUT THE COMPANY

WILDERNESS SOUTHEAST
711 Sandtown Rd, Savannah, GA 31410, (912) 897-5108

The focus of Wilderness Southeast is to foster a sense of Earth stewardship through nature study tours in the southeastern US. The nonprofit organization, in operation for 19 years, encourages guests to be active in preservation issues, while teaching them sound environmental ethics and awareness.

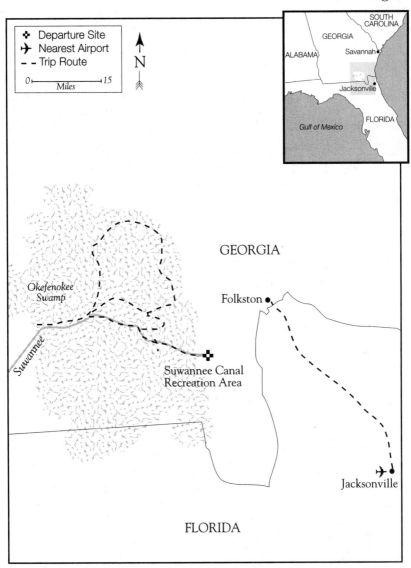

Georgia

- ❖ Departure Site
- ✈ Nearest Airport
- – – Trip Route

0 |———————| 15
Miles

N

SOUTH CAROLINA
GEORGIA
ALABAMA
Savannah
Jacksonville
FLORIDA
Gulf of Mexico

GEORGIA

Okefenokee
Swamp

Folkston ●

Suwannee

❖ Suwannee Canal
Recreation Area

✈ Jacksonville

FLORIDA

ECOFOCUS The primary problems in and around the Okefenokee Wilderness Area are the overuse of the swamp and the conflict of multiple uses of the area, such as motorboats versus canoes. Another controversy centers over whether to allow the swamp to burn during natural fires to maintain its health.

HAWAIIAN ADVENTURE
Hike, bicycle and kayak in Hawaii with EcoTours of Hawaii

Traverse from the windward to the leeward coast of Hawaii by foot and bicycle, crossing the most massive mountain in the world, Mauna Loa, estimated at 10,000 cubic miles. Paddle down the beautiful coastline in a kayak for 20 miles of incredible marine life viewing. The first seven days of the trip will entail backpacking from sea level through coastal and midland regions of Hawaii Volcanoes National Park to the summit of Kilauea Volcano. Create your own itinerary and take advantage of the opportunities for side trips, including swimming in bays and inland pools. On the eighth day of the tour, the backpacking route will lead you to a cabin snuggled in a hollow at the base of a cinder cone 10,000 feet in elevation. You will continue to head up to another cabin perched at 13,250 feet near the edge of a semi-active volcanic crater. The descent by bike will leave you coasting down to cabins in a high plain between two of the largest mountains in the world. Getting back on bicycle, you will follow a winding mountain road to the coast. Along beautiful beaches, you'll be able to relax, swim, snorkel or paddle through the crystal-blue water in a kayak. The last two days will include leisurely kayaking from one beach camp to another. Explore the mountains, volcanoes, beaches and wildlife of Hawaii as you immerse yourself in the phenomenal scenery.

ABOUT THE TRIP

LENGTH OF TRIP: 14 days

DEPARTURE DATES: June 13, Oct. 12

TOTAL COST: $1,120-$1,330

TERMS: Cash, Visa, MC; deposit of $150, 45 days before departure; cancellation policy of refund less deposit with 45 days advance notice

DEPARTURE: Arrive at Hilo International Airport on the island of Hawaii and drive in group vans to a campsite on the coast.

AGE/FITNESS REQUIREMENTS: Average physical fitness

SPECIAL SKILLS NEEDED: None

TYPE OF WEATHER: Warm in coastal areas, temperatures around 80°; cool on Mauna Loa

OPTIMAL CLOTHING: Bathing suit, layers of comfortable hiking clothes and boots

GEAR PROVIDED: All community camping gear

GEAR REQUIRED: Sleeping bag, backpack, eating utensils, sleeping pad

CONCESSIONS: Concessions may be purchased in towns.

WHAT NOT TO BRING: Radios without headphones

YOU WILL ENCOUNTER: Birds, wild goats, pigs and bountiful marine life

SPECIAL OPPORTUNITIES: Photography, fishing, special interpretive sessions on route

ECO-INTERPRETATION: Discussion of impact of development on coastal areas and marine life, along with daily environmental and natural history interpretation

RESTRICTIONS: Basic thoughtfulness toward local people is expected.

REPRESENTATIVE MENU: *Breakfast*—French toast, fresh fruit; *Lunch*—chicken, fish, veggie sandwich, chips, fruit; *Dinner*—chili, garlic bread, fruit pie; *Alcohol*—available for purchase at the supply points

ABOUT THE COMPANY

ECOTOURS OF HAWAII
P.O. Box 2193, Kamuela, HI 96743, (808) 885-7759, (800) 457-7759

EcoTours of Hawaii, in operation for two years, is a for-profit company owned by Hugh Montgomery. The staff of three specializes in nature tours throughout the Hawaiian Islands. They contribute to the preservation of historic walking trails through trail maintenance and advocacy at all levels of government. They also monitor golf course development that is encroaching on old forests. The trips, which avoid turtle habitats, employ local guides and often involve beach cleanups and other service opportunities.

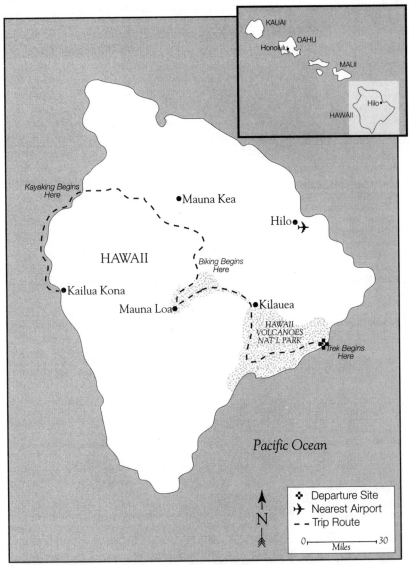

Hawaii

KAUAI

OAHU
Honolulu

MAUI

Hilo

HAWAII

Kayaking Begins
Here

●Mauna Kea

Hilo●✈

HAWAII

Biking Begins
Here

●Kailua Kona

Mauna Loa●

●Kilauea

HAWAII
VOLCANOES
NAT'L PARK

Trek Begins
Here

Pacific Ocean

N

✤ Departure Site
✈ Nearest Airport
- - Trip Route

0 ⊢————⊣ 30
Miles

ECOFOCUS Tourism development is threatening the marine life, coastline, forests and indigenous cultures of Hawaii. The shorelines are packed with looming resorts and hotels, and trail access along the beaches is limited. The historic walking trails are considered endangered, and resort golf courses are threatening the forests and ecosystems of the island.

MIDDLE FORK OF THE SALMON
Whitewater rafting in Idaho with Rocky Mountain River Tours

A clear blue sky, white sand beaches, tall green conifers and a raging Rocky Mountain river invite you to the vacation of a lifetime. The Middle Fork of the Salmon rips through 105 miles of the most primitive, remote country in the world, some of which is accessible only by the river itself. Licensed professional guides carry you and your group safely through the waves in oar-powered rafts. Inflatable kayaks and paddle boats are also available for your fun and adventure. This particular area of the Rocky Mountains is home to a wide diversity of wildlife, from enormous raptors circling overhead to majestic bighorn sheep and even an occasional mountain lion or otter. It is also rich in history. On this trip, you will also have the opportunity to visit Indian sites, including the remnants of the Tukudeka mountain tribe and the Sheepeaters tribe which was eliminated by the US cavalry in 1865. In the evening, while the staff prepares a delicious gourmet dinner, you will have time to hike up nearby canyons and gorges, try your hand at fishing (native cutthroat and rainbow trout thrive in the Middle Fork), go for a swim or just relax and enjoy the roar of the river. Rocky Mountain River Tours reveals to you the splendor of the Salmon River in pure style with minimum impact on the environment.

ABOUT THE TRIP

LENGTH OF TRIP: 4-6 days

DEPARTURE DATES: June, July, Aug.

TOTAL COST: $745-$1,145

TERMS: Check; $400 nonrefundable deposit; cancellation insurance available

DEPARTURE: Arrive at airport in Boise, ID; 2-hour drive by rental car or air taxi to the departure point

AGE/FITNESS REQUIREMENTS: All ages welcome

SPECIAL SKILLS NEEDED: None

TYPE OF WEATHER: Hot days, cool nights

OPTIMAL CLOTHING: Lightweight clothes, raingear, tennis shoes, hat and warm clothes for nighttime

GEAR PROVIDED: Everything except personal clothing

GEAR REQUIRED: Personal clothing

CONCESSIONS: Film, souvenirs and personal items can be purchased prior to trip departure in Boise.

WHAT NOT TO BRING: Tape players and phones

YOU WILL ENCOUNTER: Plenty of wildlife, including Rocky Mountain bighorn sheep, elks, deer, bears, eagles, otters and many more

SPECIAL OPPORTUNITIES: Catch-and-release fishing, hiking, paddle rafting, inflatable kayaks, soaking in hot springs

ECO-INTERPRETATION: River guides are knowledgeable in local geology, botany, biology, water ecology, ornithology and natural history.

RESTRICTIONS: Walk with respect around Indian grave sites and villages.

REPRESENTATIVE MENU: *Breakfast*—sticky buns, juice, coffee, fruit; *Lunch*—tortellini salad, sourdough bread; *Dinner*—shrimp and feta over fresh tomato linguine with sage, olive focaccia; *Alcohol*—fine wine served with dinner; beer is allowed

ABOUT THE COMPANY
ROCKY MOUNTAIN RIVER TOURS
P.O. Box 2552-BW, Boise, ID 83701, (208) 345-2400

Established in 1978, Rocky Mountain River Tours is a for-profit company specializing in whitewater rafting in Idaho. Owned by Dave and Sheila Mills, the outfitter has also stressed the importance of preserving the Salmon River and has pressured the US Forest Service into regulating campsites to insure minimal impact. Sheila is the author of *Rocky Mountain Kettle Cuisine*, which has a section on responsible outdoor cooking as well as tasty Dutch-oven recipes.

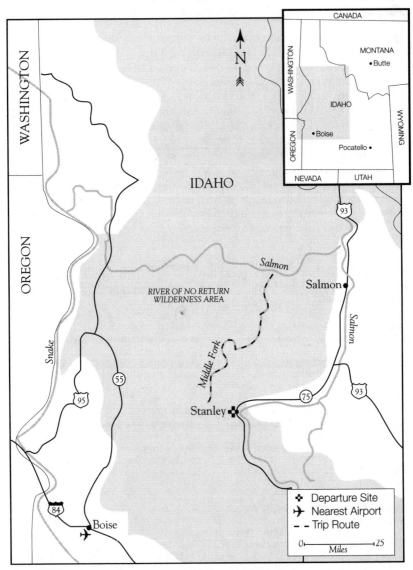

ECOFOCUS Since the Middle Fork of the Salmon is protected by Congress as a Wild and Scenic River, it is exempt from most environmental problems. Although it is extremely popular and hosts an enormous flow of visitors, the US Forest Service has instilled many valuable restrictions to limit crowds and lessen environmental impact.

RIVER OF NO RETURN STEELHEAD TRIP
Raft and fish in Idaho with Silver Cloud Expeditions

Located within the largest wilderness area in the lower 48 states, the River of No Return Steelhead Fishing Trip assures trip participants peace, serenity, solitude and, of course, fun. This unique trip covers 60 miles of the steep, wooded Salmon River Canyon, which is the second-deepest canyon in North America after the Grand Canyon. This trip is especially geared for adults looking for a small-scale, exclusive river trip with emphasis on all aspects of steelhead fishing, including their life cycle, techniques for fishing, catch-and-release philosophy and threats to future steelhead populations. The healthiest and farthest-migrating school of steelhead is found on this stretch of the river; the fish here weigh up to 20 pounds. The river runs through some big Class III rapids, offering exciting diversions to the fishing. If that isn't enough, the moose, eagles, otters, black bears, mountain sheep and other wildlife, coupled with friendly company and Dutch-oven gourmet food, round out the experience.

ABOUT THE TRIP

LENGTH OF TRIP: 4-5 days

DEPARTURE DATES: Oct. 3, 12, 21, 30, Mar. 6, 15

TOTAL COST: $685-$1,095

TERMS: Check; $200 deposit to reserve; full refund 60 days prior to departure

DEPARTURE: Nearest airport in Boise, ID; charter air transportation to departure point in Salmon, ID, provided

AGE/FITNESS REQUIREMENTS: None

SPECIAL SKILLS NEEDED: None

TYPE OF WEATHER: Cool with occasional precipitation; temperatures in October can be in the 60s in the day and below freezing at night

OPTIMAL CLOTHING: Sweater, polypropylene, fleece, waders

GEAR PROVIDED: All rafting equipment, tents and waterproof duffel

GEAR REQUIRED: Personal outdoor clothing and sleeping bag

CONCESSIONS: Film, souvenirs and personal items can be purchased in Salmon, ID, but not on the trip.

WHAT NOT TO BRING: Weapons, cellular phones and pets

YOU WILL ENCOUNTER: Excellent whitewater, wildlife and scenery

SPECIAL OPPORTUNITIES: Visits to natural hot springs

ECO-INTERPRETATION: Natural history programs are offered every night and include Salmon River fauna, fish, Indian history, catch-and-release methods and low-impact camping.

RESTRICTIONS: None

REPRESENTATIVE MENU: *Breakfast*—pancakes, eggs, fruit; *Lunch*—hearty sandwiches, soup; *Dinner*—chicken breasts, steak, fresh vegetables, dessert; *Alcohol*—wine provided; bring other alcohol if desired

ABOUT THE COMPANY

SILVER CLOUD EXPEDITIONS
P.O. Box 1006-B, Salmon, ID 83467, (208) 756-6215

Owned by Jerry and Terry Myers and in business for 16 years, Silver Cloud Expeditions is a for-profit outfitter emphasizing wilderness rafting and fishing tours on Idaho rivers. Both Jerry and Terry are very vocal when it comes to preserving river and riparian areas of the state and are active in local conservation groups, such as Idaho Rivers United.

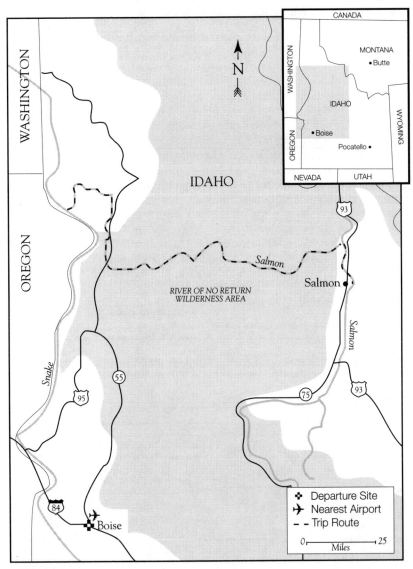

ECOFOCUS While this section of the Salmon River is protected under Wilderness Area status, salmon and steelhead runs are becoming increasingly endangered by dams, diversions and mineral and logging interests upstream and downstream, and are in need of protection. Also critical are the issues of salmon run recovery on the Columbia River, reintroduction of the gray wolf and establishing additional wilderness designation for roadless areas.

THE SEA AND LAND LIFE OF MAINE
Walk the coast of Maine with Country Walkers

Walk along the rugged shoreline scattered with harbor seals, eiders, cormorants and guillemots bobbing among colorful lobster buoys. The trip begins at Blue Hill Farm, a rustic inn located in an old shipbuilding town that is home to 75 buildings listed on the National Historic Register. An afternoon walk will take you to the Reversing Falls, mysterious rapids that flow seaward at low tide but reverse their direction as the tide rises. Another day will lead you to explore the historic seaport village of Castine, which has been claimed by four different countries since its founding in the early 17th century. The Wilson Museum and the John Perkins House, a pre-Revolution home, await your visit in the most photographed village in Maine. The last two evenings of the trip are spent at the Crocker House Inn, built in 1884. This remote inn will allow you to explore the Schoodic Peninsula and to admire the unsurpassed beauty of Mt. Desert Island. Mt. Desert is the home of more than 300 species of birds and a wide variety of sea and land wildlife. The options on the island include carriage paths around Eagle Lake, watching seals and ospreys at Indian Point, exploring tidal pools at Otter Point and hiking up Cadillac Mountain.

ABOUT THE TRIP

LENGTH OF TRIP: 5 days

DEPARTURE DATES: May 17-Oct. 30, every Sunday

TOTAL COST: May 17-June 19, $889; June 20-Oct. 30, $929

TERMS: Check, AmEx, VISA, MC; $250 deposit required to reserve; fee is 15% of $250 if cancelled at least 21 days prior to trip

DEPARTURE: Arrive in Bangor, Maine, and meet group transportation for travel from airport to Blue Hill.

AGE/FITNESS REQUIREMENTS: Minimum age of 18; general good health

SPECIAL SKILLS NEEDED: None

TYPE OF WEATHER: Temperatures ranging from 40° to 80°

OPTIMAL CLOTHING: Casual walking clothes and comfortable shoes or lightweight hiking boots

GEAR PROVIDED: None

GEAR REQUIRED: Day pack

CONCESSIONS: Concessions can be purchased on the trip.

WHAT NOT TO BRING: A list will be provided upon registration.

YOU WILL ENCOUNTER: Shorebirds, harbor seals, ospreys nesting, beaver habitats and loons

SPECIAL OPPORTUNITIES: Discussion with locals about history of fishing on coast of Maine; blueberry wine tasting at the Bartlett Maine Estate Wine Company

ECO-INTERPRETATION: Information is provided on wildflowers, ferns, nature conservancy holdings, harbor seals, nesting ospreys and acid rain.

RESTRICTIONS: None

REPRESENTATIVE MENU: *Breakfast*—juice, hot and cold cereals, eggs, pancakes, French toast; *Lunch*—picnic lunches, day at Eaton's Lobster Pier; *Dinner*—local inn specialties, lobster, soup, salad, dessert; *Alcohol*—available for purchase at inns; cannot be consumed in public areas

ABOUT THE COMPANY

COUNTRY WALKERS
P.O. Box 180 , Waterbury, VT 05676, (802) 244-1387

Country Walkers, a for-profit company owned by Bob and Cindy Maynard, has been in business for two years and has a staff of 24 people. Specializing in walking vacations in the US, New Zealand and Ireland, Country Walkers donates some of its trip proceeds to several nonprofit organizations, such as the Nature Conservancy. Using local accommodations and local specialists as guides, the Maynards are committed to environmental preservation and have revamped their bike tour company to create small-group walking trips, reducing the impact on the areas they visit.

Maine

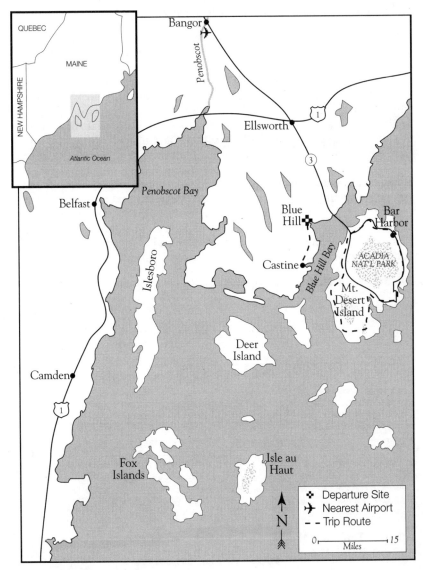

ECOFOCUS The major environmental issues confronting coastal Maine are acid rain and the adverse effects of tourism. The trees, lakes, blueberry crops and wildlife are suffering from the toxins of Northeastern industry. The toxins become increasingly debilitating as they are passed up the food chain.

THE SEA AND SKY OF MAINE
Sail the coast of Maine in the schooner Lewis R. French

Sail along the coast of Maine among the islands of Penobscot Bay, where wind, weather and whim dictate the itinerary. Passengers are encouraged and recruited to help with all aspects of sailing, from dropping the anchor to steering the 121-year-old coasting schooner. The *Lewis R. French* is the oldest schooner in the windjammer fleet, constructed of red oak and white pine. Although the route varies weekly, you will cruise along the rugged islands and shore from Boothbay to Acadia National Park on Mt. Desert, visiting picturesque fishing villages and historic towns. Twenty-three other sailors will join you on this journey, sailing through morning fog and into a horizon of brilliant stars. You will have the opportunity to partake in a traditional Maine lobster "feed," swim on a warm afternoon, make homemade ice cream, trade stories in the galley and joke on the quarterdeck. The scenic route includes bountiful sea and land wildlife such as ospreys, bald eagles, gulls, sea ducks, loons, whales, seals, porpoises and fish. Relax and settle in with a mug of hot coffee and a homemade cookie and cruise the coast of Maine with Captain Dan and Kathy Pease.

ABOUT THE TRIP

LENGTH OF TRIP: 6 days

DEPARTURE DATES: Mondays, June-Sept.

TOTAL COST: $615

TERMS: Cash, check, MC, Visa; $200 deposit; full refund if cancelled more than 28 days prior to trip

DEPARTURE: Arrive at Knox County Regional Airport, Rockland, ME, and take the coastal van service to the marina.

AGE/FITNESS REQUIREMENTS: Minimum age of 16; general good health

SPECIAL SKILLS NEEDED: None

TYPE OF WEATHER: Summer Maine weather, including sun, wind and fog; temperatures ranging from 40° to 60°

OPTIMAL CLOTHING: Casual, slacks, jeans, sweater, jacket and non-skid shoes

GEAR PROVIDED: Bedding and towels

GEAR REQUIRED: Camera, rain jacket, sunglasses, good humor and an appetite

CONCESSIONS: Souvenirs are sold on board. Film and personal items should be purchased prior to departure.

WHAT NOT TO BRING: Your job, radio or phone

YOU WILL ENCOUNTER: Local people, eagles, ospreys, gulls, sea ducks, loons, whales, seals and porpoises

SPECIAL OPPORTUNITIES: Fishing for mackerel, old-fashioned lobster cookout on an island beach, swimming

ECO-INTERPRETATION: Discussion of coastal ecosystems, sea animals and bird life

RESTRICTIONS: Delicate coastline; leave nothing on shore and do not let any trash blow overboard

REPRESENTATIVE MENU: *Breakfast*—family-style: French toast, pancakes, coffee; *Lunch*—salad, soup and sandwich; *Dinner*—roasts, fish, lobster, vegetables, homemade bread and jams; *Alcohol*—bring your own in moderation

ABOUT THE COMPANY

SCHOONER *LEWIS R. FRENCH*
P.O. Box 482 , Rockland, ME 04841, (207) 594-8007, (800) 648-4544

Schooner *Lewis R. French*, a for-profit venture owned by Captain Dan Pease, has been in business for seven years with a staff of four. The business contributes to the preservation of Maine's coastline and islands by regular coastline patrols and cleanups of camps, as well as by donating to environmental organizations.

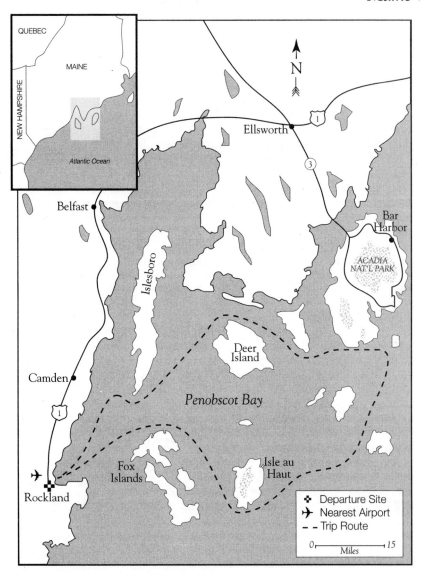

QUEBEC

MAINE

NEW HAMPSHIRE

Atlantic Ocean

Ellsworth

Belfast

Bar Harbor

ACADIA NAT'L PARK

Islesboro

Deer Island

Camden

Penobscot Bay

Fox Islands

Isle au Haut

Rockland

❖ Departure Site
✈ Nearest Airport
- - Trip Route

0 ⊢————————⊣ 15
Miles

ECOFOCUS Maine's coastline is threatened by the nearby deep-water oil shipping ports, which pollute the water and increase the chance of oil spills or leakage. Acid rain, local sewer problems and the fishing industry are depleting the once-plentiful fishery here.

ARROWHEAD BIKE/CANOE TRIP

Bicycle and canoe in the Boundary Waters with Timberline Bicycle Tours

This unique and exciting combination bike/canoe adventure is for those who love to ride and paddle amid sheer beauty. The fun kicks off in Duluth, Minnesota, with two days of cycling through the scenic North Shore Scenic Road along Lake Superior. After an overnight stay in Silver Bay, the entourage will ride into the Superior National Forest en route to an evening at the Blue Fin Lodge. From here, it's on to the canoeing portion of the trip in the Boundary Waters Wilderness for four days of peace and solitude far removed from the hustle and bustle of the civilized world. Since Boundary Waters is a protected area, the size and number of groups allowed are closely monitored and regulated. This ensures a very personal and peaceful encounter with an area that retains much of its primitive nature. The days will be spent exploring the pristine series of lakes that make up this magnificent wilderness, the nights camping right next to the water. After a return to the Blue Fin Lodge, the group will hop back on their bikes to climb the glacial ridge that forms Minnesota's Sawtooth Mountains. The ride, through the Superior Forest, is a cyclist's dream as the group winds into the heart of Minnesota's lake country on sparsely traveled roads.

ABOUT THE TRIP

LENGTH OF TRIP: 9 days

DEPARTURE DATES: July 25

TOTAL COST: $925

TERMS: Cash or check accepted; $200 deposit to reserve; full refund less $50 if cancelled 30 days prior to departure

DEPARTURE: Nearest airport in Duluth, MN; Timberline vans transfer guests to Duluth departure location

AGE/FITNESS REQUIREMENTS: Minimum age is 11; reasonably fit condition

SPECIAL SKILLS NEEDED: Cycling experience

TYPE OF WEATHER: Sunny, warm days with temperatures in the 70s; cool nights in the 50s; showers possible

OPTIMAL CLOTHING: Raingear and warm clothes for evenings

GEAR PROVIDED: Tents and cooking gear for canoe segment

GEAR REQUIRED: Sleeping bag and pad

CONCESSIONS: Film, souvenirs and personal items can be purchased along the cycling part of the trip.

YOU WILL ENCOUNTER: Wildlife, including deer, beavers, bears, numerous species of birds

SPECIAL OPPORTUNITIES: Hiking

ECO-INTERPRETATION: Guides are knowledgeable about the geology, geography, wildlife and ecology of the area. Guests are provided with interpretive material.

RESTRICTIONS: None

REPRESENTATIVE MENU: *Breakfast*—pancakes, cereal and juice; *Lunch*—salad, sandwich, fruit and beverage; *Dinner*—soup, salad, entree choice, dessert and beverage; *Alcohol*—individual responsibility

ABOUT THE COMPANY

TIMBERLINE BICYCLE TOURS
7975 E. Harvard, #J, Denver, CO 80231, (303) 759-3804
Timberline Bicycle Tours, specializing in biking/hiking/canoeing trips in the western US and northern Great Lakes, has been in business for 10 years. The for-profit company, owned by Dick and Carol Gottsegen, also lobbies and writes letters for the protection of cyclists from automobiles on highways through the national parks.

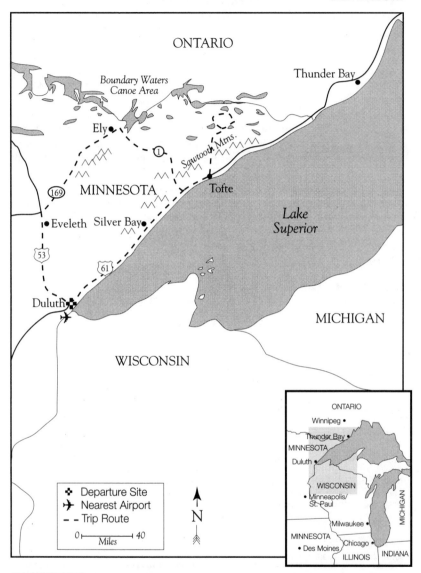

ECOFOCUS Half of the trip takes place within the highly restrictive confines of the Boundary Waters Wilderness, where adventures are arranged by a permit system so as to minimize human impact on the environment. The region is not totally without environmental problems, though, as acid rain has been observed around the eastern Great Lakes in Ontario.

PATHWAYS OF THE ANASAZI
Llama trek in New Mexico with American Wilderness Experience

The American Southwest was once home to a highly advanced civilization of indigenous peoples who, without the use of beasts of burden and even before the invention of the wheel, built an elaborate and extensive "highway system" before suddenly and mysteriously disappearing from the area over 800 years ago. The Anasazi, which translates to the "Ancient Ones," left behind many clues and insights into their lifestyles. These relics remain intact for observation by those willing to trek the arid region to see them. Using llamas to haul the gear and led by a local naturalist, you will view ancient roads as well as visit numerous native ruins and ceremonial sites, many of them still as they were when the Anasazi abandoned them. The terrain is a pleasant mix of desert grassland and ponderosa pine woodland, occasionally bisected by igneous rock formations. Enjoy endless vistas of majestic mountains, rolling plains and unforgettable sunsets, in addition to fine Southwestern cuisine, as the group learns about a highly sophisticated ancient culture whose reason for disappearing from the region remains unclear to this day.

ABOUT THE TRIP

LENGTH OF TRIP: 5 days
DEPARTURE DATES: Sept. 7, 21, Oct. 12, 19
TOTAL COST: $635
TERMS: Check, MC, VISA; $250/person deposit required within 10 days of confirmation; deposit is nonrefundable, but can be credited to another trip
DEPARTURE: The nearest airport is in Albuquerque, NM; shuttle service is available to departure point
AGE/FITNESS REQUIREMENTS: Minimum age is 12
SPECIAL SKILLS NEEDED: None
TYPE OF WEATHER: Fall weather; dry and sunny, 50s-70s
OPTIMAL CLOTHING: Loose layer method, good rainsuit and hiking boots
GEAR PROVIDED: Tents, mess gear, camp gear, sleeping bags, pads and panniers for packing gear
GEAR REQUIRED: Personal clothing, day pack and raingear

CONCESSIONS: Film should be purchased prior to trip.
WHAT NOT TO BRING: Radios, firearms, fireworks, pets
YOU WILL ENCOUNTER: Anasazi artifacts, ruins and ceremonial sites
SPECIAL OPPORTUNITIES: Stargazing and edible-plant gathering
ECO-INTERPRETATION: Back-country ethics, the art of llama packing, wildlife identification, Native American lore, legends and facts are all covered.
RESTRICTIONS: Leave all artifacts as they are found.
REPRESENTATIVE MENU: *Breakfast*—eggs, cereal, juice and coffee; *Lunch*—sandwich board, fruit and energy bars; *Dinner*—Southwestern, Tex-Mex and northern New Mexico specialties; *Alcohol*—bring your own, in moderation

ABOUT THE COMPANY

AMERICAN WILDERNESS EXPERIENCE
P.O. Box 1486, Boulder, CO 80306, (303) 444-2622, (800) 444-0099
Offering domestic and international back-country adventure travel, American Wilderness Experience has been a for-profit outfitter for the last 21 years. Owned by Dave Wiggins, it pledges to plant a tree for every participant on its trips all around the country, as well as provide each with a membership to the American Forestry Association.

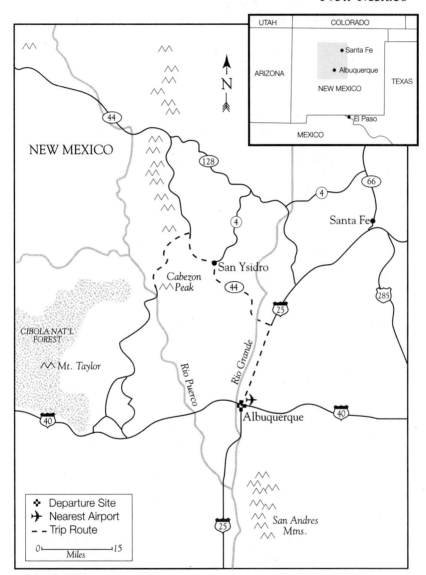

ECOFOCUS Four areas within this region of New Mexico are Wilderness Study Areas and are currently being examined in an Environmental Impact Statement. However, Wilderness designation for 800 square miles is being tied up due to disagreements over water and grazing rights. Other important issues include the illegal cutting of juniper and piñon for firewood and the protection of Anasazi ruins and other historical sites from vandals and poachers.

ANCIENT FOREST ADVENTURE
Hike Oregon with Ancient Forest Adventures

Ancient Forest Adventures has set up various trips to help people from all over the country see their endangered ancient forests while the forests remain. This trip will introduce you to every aspect of these majestic forests and to the reasons why they should be protected. The tours, conducted in five of Oregon's national forests, are designed to show the vast diversity and ecology as well as the importance of the forests as habitats. You will learn about Native American food and medicinal plants and you can participate in activities such as storytelling, drumming circles, song circles and the Native American sweat lodge, native ways of finding connections with the Earth. Trip participants will generally hike between six and eight miles per day along gradual gradients. There will be frequent stops for picture-taking, lunch and snacks. Some trips will climb higher to take advantage of the spectacular views of the Cascade Mountains and the Oregon coast. Your group will spend nights at the Breitenbush Hot Springs resort, which offers guest cabins with hot tubs, a sauna, meadow pools and organic vegetarian meals.

ABOUT THE TRIP

LENGTH OF TRIP: 6 days
DEPARTURE DATES: Apr. 26, May 17, June 21, July 12, Aug. 25, Sept. 20, Oct. 18, Nov. 15, Dec. 6
TOTAL COST: $525
TERMS: Cash, check; 20% deposit 6 weeks in advance; cancellation up to 21 days prior minus $25 fee
DEPARTURE: Arrive at airport in Redmond; further transport provided from airport or motel room to resort and trailhead
AGE/FITNESS REQUIREMENTS: Moderately fit
SPECIAL SKILLS NEEDED: None
TYPE OF WEATHER: Temperatures in the summer range from 90° days to 40° nights; spring and fall from low 80s to 30s; winter from 50° to 15°
OPTIMAL CLOTHING: Layered clothing, polypropylene underwear, waterproof boots
GEAR PROVIDED: Extra day packs, raingear and water bottles available
GEAR REQUIRED: Day pack, raingear, pile or wool sweater, hat, mittens, flashlight, whistle, toilet paper, matches, sunscreen, water bottle
CONCESSIONS: Available on an infrequent basis
WHAT NOT TO BRING: Pets, cotton clothing or jeans

YOU WILL ENCOUNTER: Indigenous people through sweat lodges and other visits; wonderful wildlife exposure including night walks to see or hear nocturnal animals such as owls, coyotes and beavers
SPECIAL OPPORTUNITIES: Participation in Native American sweat lodge, drumming circles, storytelling, yoga, meditation, healing circles, bathing in hot springs, swimming
ECO-INTERPRETATION: Interpretation is provided on Native American and contemporary medicinal and nutritional uses of forest plants, wildlife and its habitat, management issues and recent research discoveries.
RESTRICTIONS: Follow rules of Native American ceremonies when participating.
REPRESENTATIVE MENU: *Breakfast*—oatmeal, yogurt, nuts, raisins; *Lunch*—whole-wheat burritos, fruit, nuts, veggie sticks, juice; *Dinner*—organic ingredients in all meals; salad with sprouts, choice of dressings, grain, vegetarian entree, steamed vegetables, sauce, homemade bread, whole-grain dessert; *Alcohol*—none served; strongly discouraged

ABOUT THE COMPANY
ANCIENT FOREST ADVENTURES
16 NW Kansas Ave., Bend, OR 97701, (503) 385-8633, (800) 551-1043
Ancient Forest Adventures, in operation for three years, provides interpretive tours on foot, snowshoes or cross-country skis through the Pacific Northwest to raise people's consciousness about the plight of our temperate rainforests. The owner, Mary Vogel, is active as Forest Issues Coordinator for the Central Oregon Audubon Society and a member of the local Sierra Club Conservation Committee. She also will donate 5 percent of net profit to any conservation groups that work in conjunction with Ancient Forest Adventures conservation efforts.

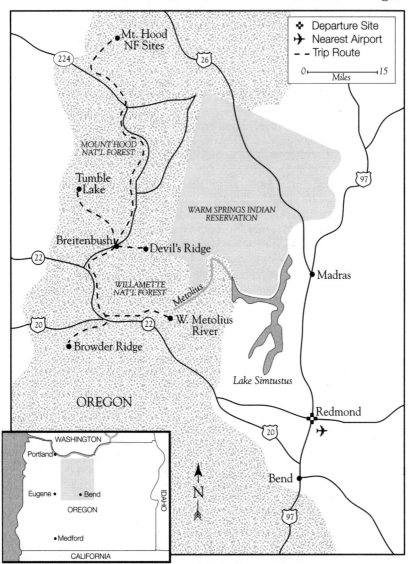

✣ Departure Site
✈ Nearest Airport
– – Trip Route

0 ⊢———————⊣ 15
Miles

Mt. Hood
NF Sites

224

26

97

*MOUNT HOOD
NAT'L FOREST*

Tumble
●Lake

*WARM SPRINGS INDIAN
RESERVATION*

Breitenbush✣ ●Devil's Ridge

22

● Madras

*WILLAMETTE
NAT'L FOREST*   *Metolius*

20   22   ●W. Metolius
River

● Browder Ridge

Lake Simtustus

OREGON

Redmond
20   ✈

WASHINGTON

Portland●

97

IDAHO

Eugene ●   ● Bend

N

Bend ●

OREGON

● Medford

97

CALIFORNIA

ECOFOCUS Logging in the Pacific Northwest has increased to such a dangerous point that species such as the spotted owl, the fisher and the northern goshawk are facing extinction due to the rapid loss of their native habitat. The 30,000 acres of ancient forest logged each year in the Northwest contribute more carbon dioxide (a greenhouse gas) to the atmosphere than all the cars in Southern California.

DIRTY DEVIL WILDERNESS TRIP
Explore remote canyons of Utah with Four Corners School

It is already warm by the time you wake up. There is sand in your bag, but you shake it out. At least it is not a scorpion! You make breakfast with the group, then start hiking. The trail winds around red sandstone spires and over mounds of slickrock. Tiny lizards scamper through the cactus. The rocks are an astounding array of reds, oranges, purples and greens. Your trail dips down to the bottom of the canyon, then winds around and through a shallow stream. As you prepare dinner, you watch shapes form in the shadows on the cliffs around you. Then you drift off to sleep under a blanket of stars. You will spend six days exploring the remote and rarely visited canyons of the Dirty Devil River, located east of Hanksville, Utah. This 500-square-mile canyon system has inspired proposals for National Park as well as Wild and Scenic River status. It is threatened by potential development of tar sand, uranium, oil and gas. This glorious area of the Southwest has scenic, recreational and bioregional integrity from canyon rim to canyon rim. A 1990 FCS participant, Herman P. Sandford, observed, "We touched it, loved it and know it better than we did. And I expect some in our group have already accepted the idea that it is their vocation to heal the Earth."

ABOUT THE TRIP

LENGTH OF TRIP: 8 days (6 days backpacking, 2 days traveling)

DEPARTURE DATES: May 18

TOTAL COST: $475

TERMS: VISA, MC, checks; $100 nonrefundable deposit; cancellation up to 60 days prior

DEPARTURE: Nearest airport is in Cortez, CO; $10 pickup fee each way to Monticello

AGE/FITNESS REQUIREMENTS: All ages; good physical condition

SPECIAL SKILLS NEEDED: Backpacking

TYPE OF WEATHER: Hot to cool

OPTIMAL CLOTHING: Medium boots, medium-weight to heavy clothing

GEAR PROVIDED: Tents and sleeping bags to rent, group gear is provided

GEAR REQUIRED: Backpack, sleeping bag, sleeping pad

CONCESSIONS: Personal items and film should be purchased in advance.

WHAT NOT TO BRING: Any excess clothing

YOU WILL ENCOUNTER: Wildlife, including bighorn sheep and deer

SPECIAL OPPORTUNITIES: Great photography

ECO-INTERPRETATION: Interpretation of local geology, natural history and geology is provided throughout by guides.

RESTRICTIONS: Limited numbers of people allowed in area

REPRESENTATIVE MENU: *Breakfast*—cereal, powdered milk, fruit; *Lunch*—crackers, cheese, fruit, salami; *Dinner*—rice, vegetables, tortillas, dessert; *Alcohol*—bring your own

ABOUT THE COMPANY
FOUR CORNERS SCHOOL OF OUTDOOR EDUCATION
East Route, Monticello, UT 84535, (801) 587-2156

In business eight years, Four Corners School of Outdoor Education is a nonprofit organization providing educational and environmental opportunities throughout the Colorado Plateau. It strives to increase awareness and sensitivity to the physical and cultural heritage of this rich and varied environment. For the past three years, it has worked with the Southern Utah Wilderness Alliance, offering Wilderness Advocacy Scholarships that focus on preservation of the Colorado Plateau.

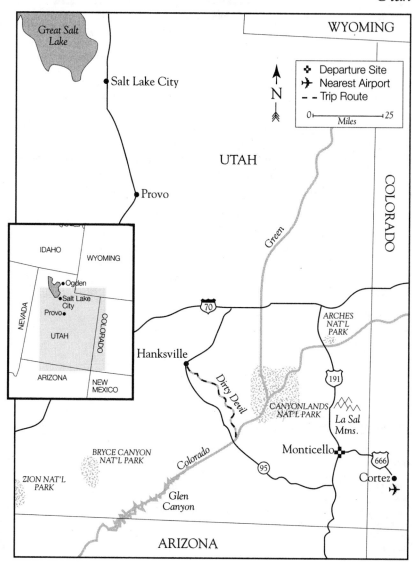

ECOFOCUS This is a proposed Wilderness Study Area of the Utah Wilderness Coalition. It is endangered because it is not included in the BLM Wilderness Proposal bill now before Congress. The area is in danger of being used for grazing cattle, as well as oil, gas and tar sand development.

ENDANGERED SPECIES OF WESTWATER
Raft the Colorado River with Canyonlands Field Institute

Hop aboard! This raft is going on a wild ride, and you're invited. You say you've never seen water any other color than blue? Join expert river runners from Canyonlands Field Institute on this stretch of the Colorado River between western Colorado and eastern Utah, where flat stretches of the red, silty water that gave the Colorado its name alternate with raging white rapids. But that's not all. This special program combines the opportunity to study the ecology of this unique canyon and to experience the thrill of running its rapids. Westwater Canyon, aside from being a scenic jewel on the Colorado River, is home to several endangered species of fish, including the humpback chub and the Colorado squawfish, as well as bald eagles and peregrine falcons. River otters, desert bighorn sheep, golden eagles and great blue herons also call Westwater home. Throughout this three-day journey, professional wildlife and fishery biologists, including Rich Valdez of Bio-West, Ana Dronkert-Egnew from the US Forest Service in Utah and Miles Moretti, a Utah Division of Wildlife Resources nongame biologist, provide in-depth discussion of water development and politics, endangered fish and other wildlife and wildlife, management in the Colorado River corridor.

ABOUT THE TRIP
LENGTH OF TRIP: 3 days
DEPARTURE DATES: July 10
TOTAL COST: $350
TERMS: Cash, checks, charge; $125 deposit; cancellation up to 45 days prior
DEPARTURE: Nearest airport in Grand Junction, CO; rental cars and cabs available for transportation to Moab
AGE/FITNESS REQUIREMENTS: Must be over 16 years old
SPECIAL SKILLS NEEDED: None
TYPE OF WEATHER: Very warm during the days, in the 90s; cool at night, in the 50s
OPTIMAL CLOTHING: Swimsuit, both light- and heavyweight clothes, sneakers or river sandals, raingear, hat, sunglasses
GEAR PROVIDED: Boating equipment
GEAR REQUIRED: Overnight camping equipment, fishing gear (optional)

CONCESSIONS: Bring film and personal items with you.
WHAT NOT TO BRING: Excess baggage
YOU WILL ENCOUNTER: Extensive wildlife including humpback chub, Colorado squawfish, bald eagles, peregrine falcons river otters, desert bighorn sheep, golden eagles and great blue herons
SPECIAL OPPORTUNITIES: Fishing
ECO-INTERPRETATION: This trip is led by highly acclaimed regional biologists who will provide extensive ecological interpretation.
RESTRICTIONS: Guides will discuss river etiquette.
REPRESENTATIVE MENU: *Breakfast*—pancakes, eggs, fruit; *Lunch*—bread, fruit, vegetables, cheese, meats; *Dinner*—meat, vegetables, bread, dessert; *Alcohol*—not provided

ABOUT THE COMPANY
CANYONLANDS FIELD INSTITUTE
P.O. Box 68, Moab, UT 84532, (801) 259-7750
Dedicated to promoting understanding, appreciation, and stewardship of the natural environment and cultural heritage of the Colorado Plateau, CFI is a nonprofit organization in business eight years. It employs local specialists to lead trips and brings groups to the Navajo and Hopi reservations to work with native guides. CFI is involved with the local school system. It also works with Southwestern outfitters to promote desert etiquette and help establish educational dialogue between a variety of local groups, including farmers, miners and outfitters.

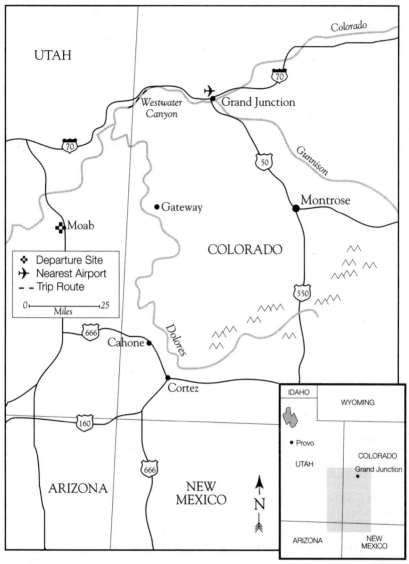

ECOFOCUS Endangered fish, a result of polluted streams and the introduction of damaging fish species, are an increasing environmental problem in Westwater Canyon. Powerboats and jet skis on the upper sections of the river also threaten and scare fish, nesting birds and deer. Both of these problems stem from the increased use of the area by recreationists.

SOUTHERN UTAH ODYSSEY
Raft, mountain bike and hike in Utah with High Desert Adventures

During this odyssey, trip participants will discover a panorama of deserts, mountains and canyons in five of Utah's national parks: Arches, Canyonlands, Capitol Reef, Bryce and Zion. Private vehicles transport guests from park to park, but mountain bikes, rafts and your own feet provide the rest. Participants start this outdoor adventure with hiking in Arches around many of its sandstone fins and arches, then climbing one of the 12,000-foot peaks in the nearby Sierra La Sal Mountains. Guests then descend 7,000 feet on mountain bikes on their way to Canyonlands. The biking portion continues from the lofty Island in the Sky massif down to the mighty Colorado River, where everyone boards rubber rafts to take on the lively rapids of Cataract Canyon. Anasazi petroglyphs await your discovery in the vast Waterpocket Fold of Capitol Reef when the group mounts the bikes again for another descent down to the Calf Creek oasis of the rugged Escalante Wilderness. Bryce Canyon is the location of magnificent scenery of limestone formations as the group hikes in the company of thousands of pinnacles, spires and hoodoos, which are human-like sandstone formations. The odyssey concludes in Zion, where peregrine falcons soar high above as you admire the sheer 2,000-foot sandstone cliffs at the Narrows of the Virgin River.

ABOUT THE TRIP

LENGTH OF TRIP: 11 days
DEPARTURE DATES: Apr.-Oct.
TOTAL COST: $1,495
TERMS: Check, credit card, money order; $300 deposit required to reserve and final payment 60 days prior; full refund less $50 if cancelled 60 days or more before departure
DEPARTURE: Nearest airport in Salt Lake City, UT; hotel airport shuttle provides transport to Salt Lake City departure point
AGE/FITNESS REQUIREMENTS: None
SPECIAL SKILLS NEEDED: None
TYPE OF WEATHER: Warm, hot and dry in desert with temperatures between 80° and 100°; cooler in mountains
OPTIMAL CLOTHING: Shorts, T-shirt and light-weight hiking boots
GEAR PROVIDED: Rafting, biking, kitchen and eating gear
GEAR REQUIRED: Personal clothing and camping gear (camping gear rental available)

CONCESSIONS: Film, souvenirs and personal items can be purchased along the way.
WHAT NOT TO BRING: Firearms, radios and pets
YOU WILL ENCOUNTER: Spectacular canyon and mountain scenery
SPECIAL OPPORTUNITIES: View Indian ruins and swim in the rivers of the La Sal mountain range.
ECO-INTERPRETATION: Talks on current environmental issues in the region, local geology and pre-Columbian Indian culture are provided throughout the trip.
RESTRICTIONS: No collecting at archaeological sites, low-impact camping
REPRESENTATIVE MENU: *Breakfast*—pancakes, eggs, yogurt and potatoes; *Lunch*—pasta salads, sandwiches and fruit; *Dinner*—Dutch-oven specialties; pasta, Mexican salads and desserts; *Alcohol*—guests bring their own

ABOUT THE COMPANY

HIGH DESERT ADVENTURES
757 E. South Temple, Ste. 201, Salt Lake City, UT 84102, (800) 345-RAFT
Running outdoor trips to Idaho and the American Southwest for the last nine years, High Desert Adventures attempts to instill an appreciation for the natural history and culture of a region. This for-profit outfitter also supports the Grand Canyon Trust by featuring it in its catalog and through mailings to trip guests. It also supports the Southern Utah Wilderness Alliance.

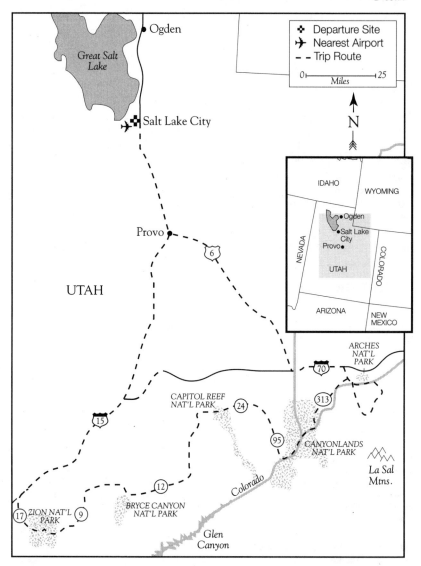

ECOFOCUS Southern Utah is a hotbed of environmental dispute. The battle is between those who want to introduce and expand logging, mining and power plant operations in the area in an attempt to boost the local economy, and groups who want to preserve the precious natural resources in this area.

WASATCH WILDERNESS
Back-country skiing in Utah with Powder Ridge Ski Touring

Nestled in a remote corner of northern Utah's Wasatch Mountains, this back-country ski tour introduces the nordic skier to incredible powder, stunning scenery and unique winter camping in Powder Ridge's own yurts. Located adjacent to Mt. Naomi wilderness, this tour is an invitation to a wide variety of terrains ranging from gentle meadows to extreme powder chutes. Whatever your skiing taste, this tour can satisfy you. Besides the incredible skiing, one of the memorable highlights is spending your evenings in yurts, which are round, canvas-covered structures, 16 feet in diameter, that provide warm, comfortable winter lodging even in the most severe conditions. Used for centuries by nomadic peoples in Central Asia, yurts are a welcome refuge from long days of skiing and exploring. Redwood-lined skylights allow clear visibility of winter constellations. The choice is yours—to embark on a back-country trip with or without a guide. If you're not completely confident of your compass-reading skills or your avalanche-safety and first-aid techniques, then let Powder Ridge provide you with one of its experienced guides.

ABOUT THE TRIP

LENGTH OF TRIP: Custom-designed, ranging from 2 to 5 days

DEPARTURE DATES: Mid-Dec.-Mar.

TOTAL COST: Prices vary with length and type of trip.

TERMS: Check accepted form of payment; full payment for reservation; 50% refund if cancelled up to 14 days prior

DEPARTURE: Meet at trailhead outside of Logan, UT

AGE/FITNESS REQUIREMENTS: No age requirements; very good condition

SPECIAL SKILLS NEEDED: Must be able to ski with a pack

TYPE OF WEATHER: Sun, snow and wind; temperatures range from sub-0° to 32°

OPTIMAL CLOTHING: Water-repellent snow pants and jacket, warm hat and gloves

GEAR PROVIDED: All cooking, eating and yurt supplies

GEAR REQUIRED: Skis, backpack and personal gear

CONCESSIONS: Film, souvenirs and personal items can be purchased in Logan, UT, but not along route.

WHAT NOT TO BRING: Pets

YOU WILL ENCOUNTER: Terrific snow conditions and occasional small and big game

SPECIAL OPPORTUNITIES: Low-level mountaineering

ECO-INTERPRETATION: The leader of the trip is very knowledgeable about the local flora and fauna.

RESTRICTIONS: None

REPRESENTATIVE MENU: *Breakfast*—pancakes, sausage, omelettes, fruit dishes and quiche; *Lunch*—fruit and cheeses, sandwiches, hot chocolate and juice; *Dinner*—pasta, potato dishes, wild game, dessert; *Alcohol*—no restrictions

ABOUT THE COMPANY

POWDER RIDGE SKI TOURING
7124 W. Hwy 30, Petersboro, UT 84325, (801) 752-9610
Ski trips into the back-country of northern Utah are the specialty of Powder Ridge Ski Tours, which has been operating for eight years. This for-profit outfitter is owned by Ken Guest, who boasts a master's degree in environmental and outdoor education and has taught environmental education workshops for the Forest Service.

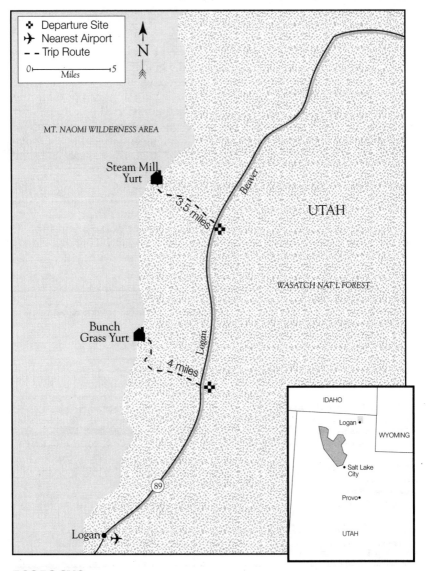

Departure Site
Nearest Airport
Trip Route

0 ━━━━━━━━━━━ 5
Miles

N

MT. NAOMI WILDERNESS AREA

Steam Mill
Yurt

3.5 miles

Beaver

UTAH

WASATCH NAT'L FOREST

Bunch
Grass Yurt

4 miles

Logan

89

Logan

IDAHO

Logan •

WYOMING

• Salt Lake
City

Provo•

UTAH

ECOFOCUS The primary environmental concerns in this area include keeping sheep and cattle out of the Wilderness Area and the ongoing fight to save scenic Logan Canyon from having its highway widened. Also, the Mt. Naomi Wilderness is the only Wilderness Area in the country that still allows the use of helicopters for killing coyotes.

BEST OF THE NORTHWEST
Hike, raft and sea kayak in Washington with Adventure Associates

This hike-raft-sea-kayak trip takes guests through three distinct ecosystems that truly showcase some of the best the Northwest has to offer. From the rugged snowcapped peaks, into the grandiose old growth forests, to the unique beauty of the pristine islands, the outdoor enthusiast will never forget this excursion. The trip begins amid the jagged peaks of the North Cascades mountains, which are sometimes referred to as the North American Alps. For the first two days, camp is set in old growth forests that also serve as the base for day hikes along mountain and forest trails. On the fourth day, trip participants will meet their river guides and start down the Class III-IV Sauk River for some wild whitewater mixed with mellow stretches. After an evening spent at a rustic inn in the town of La Conner, the group boards a ferry for the ride through the San Juans to Orcas Island for the sea kayaking leg, the final portion of the trip. The currents, tide and weather will determine the exact destination for the evening, as guests paddle through the picturesque islands famous for their rich marine life. Porpoises, eagles, harbor seals, sea lions and killer whales are all found in this area.

ABOUT THE TRIP

LENGTH OF TRIP: 9 days

DEPARTURE DATES: July 31-Aug. 8

TOTAL COST: $1,195

TERMS: Check; $200 deposit; $100 nonrefundable to cancel

DEPARTURE: Nearest airport in Seattle; Adventure Associates arranges ground transport to departure point

AGE/FITNESS REQUIREMENTS: No age requirements; average health

SPECIAL SKILLS NEEDED: None

TYPE OF WEATHER: Mild, dry and sunny with temperatures in the mid-70s in the day and 60s at night

OPTIMAL CLOTHING: Sturdy boots and backcountry travel clothing

GEAR PROVIDED: All specialized outdoor and camping equipment

GEAR REQUIRED: Personal clothing and a soft duffel bag

CONCESSIONS: Film, souvenirs and personal items can be purchased along the way.

YOU WILL ENCOUNTER: Diverse ecosystems and extensive wildlife

ECO-INTERPRETATION: Structured discussions about minimal-impact camping and conservation practices appropriate for various ecosystems

RESTRICTIONS: None

REPRESENTATIVE MENU: *Breakfast*—eggs, cereals, fruit and breads; *Lunch*—salads, fish, stews, fruits and breads; *Dinner*—fish, chicken, vegetables, pasta, pizza; *Alcohol*—discouraged

ABOUT THE COMPANY

ADVENTURE ASSOCIATES
P.O. Box 16304, Seattle, WA 98116, (206) 932-8352

In operation for four years, Adventure Associates leads cross-cultural and wilderness ecotours to the Pacific Northwest, Africa and Central America. This for-profit company, owned by Chris Miller and Sandy Braun, also makes contributions to organizations that support ecotourism, such as the Association of Experiential Education and the African Wildlife Foundation.

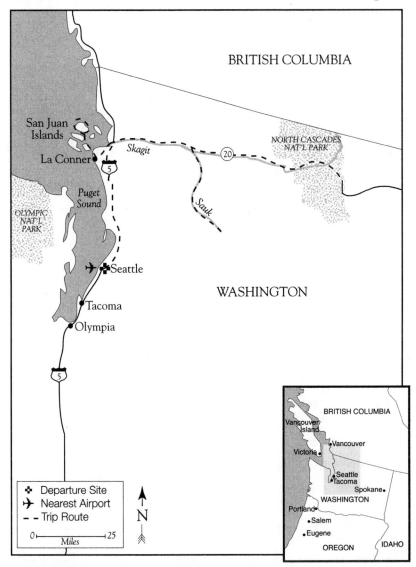

ECOFOCUS The loss of critical wildlife habitat in the Pacific Northwest, coupled with the rapid deforestation of the area's old growth forests, continue to be the primary environmental issues facing the region. Other problems include the challenge of managing users of the wild areas so as to minimize their conflicts and physical impact, as well as the controversy of whether to let oil tankers in the Puget Sound area.

OUTDOOR LEADERSHIP IN THE CASCADES
Backpack and climb in Washington with Outdoor Discoveries

This unique and challenging adventure is designed for people pursuing a career in outdoor education, interested in creating an outdoor program or hoping to expand their skills as an outdoor leader. The course is within the Wenatchee National Forest, located in the central and North Cascades of Washington state, consisting of wilderness, high alpine meadows and lakes, and forests of Douglas fir and ponderosa pine. The trip orientation includes a two-day ropes course to develop trust, communication, teamwork and cooperation. During the third and fourth days, instruction will be given at base camp and will include discussions of meal and nutritional planning and how to select and purchase food items. Instruction will be offered on the technical, interpersonal and procedural aspects of leadership. The core of the trip will take place over the next 15 days, centering on field-based instruction in leadership skills required to backpack 40 to 60 miles, ascend peaks, complete a one- to two-day solo, rock climb and rappel. The course will culminate in participants designing their own outdoor adventure course, and a group and individual evaluation of leadership goals and skills.

ABOUT THE TRIP
LENGTH OF TRIP: 22 days

DEPARTURE DATE: June 28

TOTAL COST: $1,490

TERMS: Check, MC, VISA; deposit of $373; cancellation fee of $100 if at least 30 days' notice is given; no refunds within 9 days of trip

DEPARTURE: Arrive in Seattle and drive with the group to Monroe, WA, where course begins.

AGE/FITNESS REQUIREMENTS: Minimum age of 19; strong physical condition

SPECIAL SKILLS NEEDED: Wilderness experience

TYPE OF WEATHER: Daytime temperatures 70°-80° with nighttime temperatures 30°-50°

OPTIMAL CLOTHING: Comfortable layers, hiking boots, camping clothes

GEAR PROVIDED: All group equipment

GEAR REQUIRED: Raingear, backpack, sleeping bag and pad

CONCESSIONS: Concessions can be purchased in Seattle or Monroe, but not during trip.

WHAT NOT TO BRING: Hard luggage

YOU WILL ENCOUNTER: Incredible mountain scenery

SPECIAL OPPORTUNITIES: Fishing

ECO-INTERPRETATION: Trip participants will be taught minimum-impact camping skills.

RESTRICTIONS: None

REPRESENTATIVE MENU: *Breakfast*—wholewheat, gingerbread or corn pancakes, hash browns with cheese, hot cereal, cold cereal, dried fruit, omelettes and egg dishes, hot and cold drinks; *Lunch*—cheese, cream cheese, salami, peanut butter and jelly on bagels, English muffins or pita bread, apples, oranges, nuts and dried fruit; *Dinner*—pasta with tomato sauce, peanut sauce, chicken or shrimp, macaroni and cheese, curried rice, vegetables and chicken sauce, enchiladas, soup, stew and dessert; *Alcohol*—not permitted

ABOUT THE COMPANY
OUTDOOR DISCOVERIES
P.O. Box 7687, Tacoma, WA 98407, (206) 759-6555
Outdoor Discoveries is a for-profit company that has been in business for three years, specializing in educational adventures in Scotland and the northwestern and southwestern US. A staff of five helps people learn about themselves, other cultures and the environment through minimum-impact back-country travel. The company is owned by Bob Stremba.

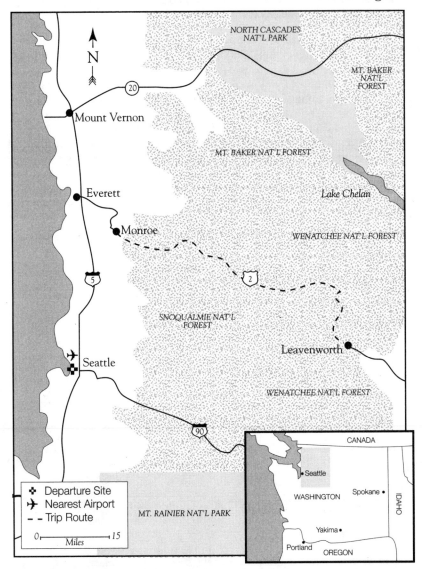

ECOFOCUS Heavy logging and the subsequent loss of wildlife habitat in Washington and the rest of the Pacific Northwest continues to be a pressing environmental issue. Also important are the problems of acid rain, the overuse of trails and the devastation of public lands by livestock.

READING THE WOLF RIVER RAPIDS
Kayak and canoe Wisconsin's Wolf River with Whitewater Specialty

You will be thrilled as you paddle your canoe or kayak past bald eagles, great blue herons, otters and other wildlife on this three-day adventure on the Wolf River. Learn to paddle safely, skillfully and confidently as you break down complex rapids into manageable parts using the paddling technique, river-reading skill and sensible judgment you will learn. Take control on the water as the river gradually progresses from a meandering pace to pushy and demanding rapids the farther downstream you go. As you gain confidence and control in progressively more difficult situations, instructors help you understand more about the river ecosystem, threats to the watershed from mining interests upstream, the impact of logging along the river and how the Menominee Indian tribe has protected the river. Intermediate and advanced classes paddle in the Menominee Reservation, where rapids like Ducknest and Big Smoky Falls offer greater challenges and inspire greater respect and understanding of the river.

ABOUT THE TRIP

LENGTH OF TRIP: 3 days (day trips)

DEPARTURE DATES: Every Friday from Memorial Day through Labor Day

TOTAL COST: $275

TERMS: MC, VISA, checks; 50% down; 100% refund for cancellation 30 days before trip

DEPARTURE: Nearest airport is Antigo, WI. Transportation made by arrangement from Antigo to Langlade.

AGE/FITNESS REQUIREMENTS: Good physical condition

SPECIAL SKILLS NEEDED: Comfortable in water

TYPE OF WEATHER: Changeable

OPTIMAL CLOTHING: Clothes to be wet in, clothes to be dry in and secure footwear

GEAR PROVIDED: Boats, paddles, personal flotation devices, helmets, wet suits, sprayskirts, rescue gear

GEAR REQUIRED: Footwear that stays on while swimming

CONCESSIONS: Film, souvenirs and personal items can be purchased at company shop.

WHAT NOT TO BRING: Alcohol, drugs, cans or bottles on the water

YOU WILL ENCOUNTER: The Menominee Indian Reservation, bald eagles, ospreys, great blue herons, minks, otters, beavers, deer, occasional bears and bobcats

SPECIAL OPPORTUNITIES: Numerous photo opportunities, trout fishing, mountain biking

ECO-INTERPRETATION: Discussions of prevalent species in the area and the impact of logging on the river's ecosystem

RESTRICTIONS: Trespass permits are required to access the river in the reservation.

REPRESENTATIVE MENU: *Breakfast*—not provided; *Lunch*—sandwiches, raw veggies, fresh fruit, granola bars, cookies; *Dinner*—fried chicken, dressing, mashed potatoes and gravy, salad, green beans, cake; *Alcohol*—not on the water

ABOUT THE COMPANY

WHITEWATER SPECIALTY
N3894 Hwy. 55, White Lake, WI 54491, (715) 882-5400
A for-profit company in business for 15 years, Whitewater Specialty has a staff of 20, teaching kayaking and canoeing students to paddle skillfully and safely down the Wolf River in Wisconsin. Owned by Bill and Donna Kallner, the company focuses on developing students' understanding and appreciation of water resources, including the effects of upstream mining.

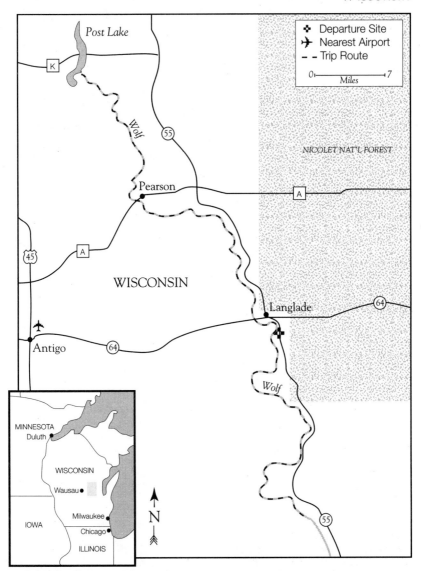

ECOFOCUS Exploring the Wolf River brings up questions of domestic water-resource management. Upriver from the Menominee Indian Reservation is an area in which mining and logging interests prevail, and the threat to the Wolf River is present as the student learns how to read and respect the river.

WILDERNESS NATURAL HISTORY COURSE
Learn wilderness skills in Wyoming with NOLS

If what you're seeking is an experience that is vigorous, challenging and fun, NOLS Wilderness Natural History Course is for you. If you enter the course with an open mind and try your hardest, you will come away with strength and a new awareness about yourself and the natural world. Your month will be spent in the rugged Absaroka and Beartooth ranges. This beautiful, isolated place can be harsh and commands respect. The hiking is challenging, the weather and terrain are demanding and the rewards will be outstanding. You will learn basic outdoor living skills, including cooking, staying warm, fire building, sanitation and stove use. You will learn knots and rope handling needed for crossing wide mountain streams or snow-fields. Orientation will also be taught using maps and compasses. Ornithology, glaciology, meteorology and identification of plants and animals will also be offered. Emphasis will be placed specifically on the natural history of the Rocky Mountain region. Through outdoor classes, active participation, journals and field reports, people dedicated to learning about the outdoors and ecology will find this trip fascinating. For those looking for total wilderness immersion and a chance to gain self-confidence, consider this a successful end to your search.

ABOUT THE TRIP

LENGTH OF TRIP: 30 days
DEPARTURE DATES: July 6, 30
TOTAL COST: $2,300
TERMS: Check, major credit card; $150 registration fee
DEPARTURE: Shuttle available from Riverton, WY; departure point is Lander, WY
AGE/FITNESS REQUIREMENTS: Must be at least 16 years old; in good shape
SPECIAL SKILLS NEEDED: None
TYPE OF WEATHER: Quick-changing mountain weather; warm days ranging from 70° to 80°, cool nights around 40°
OPTIMAL CLOTHING: Detailed list provided
GEAR PROVIDED: Gear available for rent

GEAR REQUIRED: Polypropylene underwear, wool socks
CONCESSIONS: Concessions can be purchased in Lander.
YOU WILL ENCOUNTER: Lots of wildlife, including elks, moose and bighorn sheep
SPECIAL OPPORTUNITIES: Fishing
ECO-INTERPRETATION: Teaching is done both formally and informally. Ornithology, meteorology, glaciology, and plant and animal identification are emphasized.
RESTRICTIONS: None
MEALS: NOLS students prepare their own meals from rations that include rice, pasta, beans, flour, cereals, etc.

ABOUT THE COMPANY

NATIONAL OUTDOOR LEADERSHIP SCHOOL
P.O. Box AA, Dept. ET, Lander, WY 82520, (307) 332-6973
NOLS is a nonprofit outdoor-education program that has been teaching wilderness skills and leadership for 26 years. Trips are offered in North America, Mexico, Kenya and Chile. NOLS's core curriculum includes ecology, environmental studies and land management with an emphasis on minimum-impact camping. All waste is carried out, and students achieve self-reliance and a sense of teamwork.

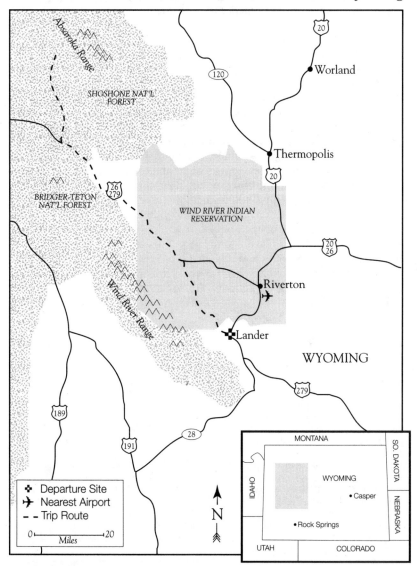

ECOFOCUS The Greater Yellowstone area is the last stronghold for the grizzly bear in the lower 48 states, but is currently facing a perilous future. The root of the problem is shrinking habitat due primarily to road building for logging, oil and gas exploration and mining interests. Also, a gold, silver and copper mine is being proposed and would be located a scant two miles from the borders of Yellowstone National Park. This would further impact grizzly habitat as well as substantially affect water drainages into the park.

PHOTO SAFARI AND WILDLIFE RESEARCH
View 60 species in five days with Great Plains Wildlife Institute

Surround yourself with Great Plains wildlife on this photo safari and wildlife research excursion. During the five-day extravaganza you'll find yourself in first-rate hotels, rustic lodges and local ranches, as Great Plains Wildlife Institute takes you from scenic Jackson Hole, Wyoming, throughout northwestern Wyoming, starting with a gentle float down the Snake River to see otters and bald eagles. The second day you'll explore Grand Teton National Park, viewing pronghorn antelope and mule deer as they graze on the rocky slopes and mountain meadows. The third day will take you to Yellowstone's geysers, buffalo and bears, and you'll visit a mountain lion researcher before you rest for the night at a secluded ranch in the Sunlight Basin with the moose and coyotes. The fourth day sets you into the middle of "The Big Empty" to photograph and document distribution of wild horses and feel the hush of the badlands, after a morning at the Buffalo Bill Cody Museum discussing feral-horse ecology, adoption programs and politics. On the fifth day you'll go to the prairies to ear-tag prairie dogs and to search for golden eagles and other birds of prey to document their nesting activity and productivity, then take a sunset drive back through Yellowstone and Grand Teton's splendors. All along, you'll be guided by expert wildlife biologists who will provide insight about the wildlife you see and the environmental problems they face.

ABOUT THE TRIP

LENGTH OF TRIP: 6 nights and 5 days; 1-day trips also available

DEPARTURE DATES: 15 trips in summer, 7 in winter

TOTAL COST: $1,635

TERMS: Cash, check, MC, VISA; $350 deposit, balance 30 days before trip; cancellations 30 days prior to trip, $350 fee, 30-15 days prior, 50% refund; 14-0 days prior, no refund

DEPARTURE: Nearest airport is Jackson, WY; transportation from airport to departure point is included

AGE/FITNESS REQUIREMENTS: 18 and over

SPECIAL SKILLS NEEDED: None

TYPE OF WEATHER: 85-90° days, 40° nights

OPTIMAL CLOTHING: Casual clothing, layers, hiking shoes

GEAR PROVIDED: Spotting scopes, binoculars, video camera

GEAR REQUIRED: Clothing, camera

CONCESSIONS: Film, personal items and souvenirs can be purchased in towns.

WHAT NOT TO BRING: Sandals

YOU WILL ENCOUNTER: 60 species of wildlife, including moose, ospreys, bald eagles, elks, antelope, white pelicans, river otters, yellow-bellied marmots, buffalo, wild horses, bighorn sheep, coyotes, pronghorn antelope and golden eagles

SPECIAL OPPORTUNITIES: Photography, fishing, hands-on research

ECO-INTERPRETATION: Expert biologists and guest lecturers guide and interpret, but participants also actively contribute to research projects.

RESTRICTIONS: Experts in animal behavior decide when it's appropriate to go in close to see the animals.

MEALS: Provided in a restaurant setting

ABOUT THE COMPANY

GREAT PLAINS WILDLIFE INSTITUTE
P.O. Box 7580, Jackson Hole, WY 83001, (307) 733-2623

Great Plains Wildlife Institute is a for-profit company that has been in business for six years and employs a staff of five. Owned by Tom Segerstrom, it specializes in low-impact wildlife viewing in northwestern Wyoming. Great Plains Wildlife Institute works with expert biologists, making observations and collecting data to aid resource-management agencies in making decisions.

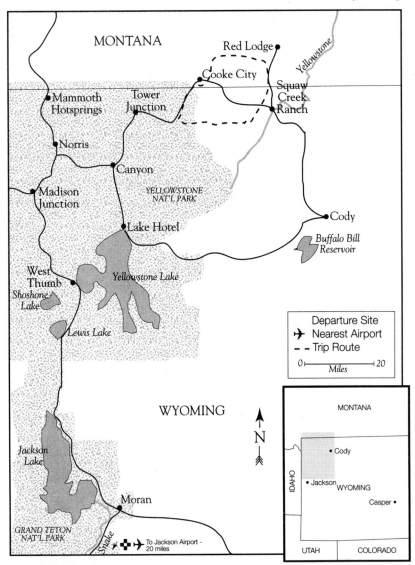

ECOFOCUS Increased tourism and the strain it puts on the natural resources are important issues in this region, especially in Yellowstone National Park and the neighboring town of Jackson. Also, logging, mining, oil exploration and ranching have a great impact on Wyoming's wild areas. These types of developments have caused habitat loss, and the grizzly bear, peregrine falcon, black-footed ferret, bald eagle and other animals have become threatened or endangered.

SEA KAYAKING THE CLAYOQUOT SOUND

Sea kayak in British Columbia with Tofino Expeditions

This exciting excursion begins from the fishing village of Tofino, British Columbia, on Vancouver Island, as the group paddles their sea kayaks past majestic Meares Island, where some of the world's largest red cedars can be found. The first few days are spent exploring the shorelines and forests of Clayoquot Sound, where every tide reveals new treasures of spectacular scenery; the presence of wildlife is breathtakingly apparent. Eagles soar high above surveying their domain, as curious porpoises playfully follow the kayaks far below. Perhaps the highlight of wildlife watching in the Clayoquot Sound, though, is provided by the resident population of gray whales, who find the food-rich habitat of the area much to their liking. Kayakers watch with exhilaration as the whales dive and feed close by in the open bays. Afternoons of hiking and beach walking offer a welcome respite from the water and allow guests to visit an ancient native village site and experience a traditional meal served at a contemporary native village, giving insight into modern native society. The overall pace of the trip is one of leisure, as time is taken to fully enjoy the many facets of this unique meeting of mountain and water.

ABOUT THE TRIP

LENGTH OF TRIP: 6 days

DEPARTURE DATES: July-Sept.

TOTAL COST: $630

TERMS: Check, VISA, MC; $200 deposit 60 days prior to departure; full refund 60 days or more, less $75 handling fee

DEPARTURE: Arrive at Vancouver International; take floatplane or bus to Tofino, BC, departure point.

AGE/FITNESS REQUIREMENTS: Ages 15 and up; good health

SPECIAL SKILLS NEEDED: None

TYPE OF WEATHER: Variable coastal weather, 55°-75°

OPTIMAL CLOTHING: Synthetic fabric, quick-drying clothing

GEAR PROVIDED: All kayaking, camping and safety equipment

GEAR REQUIRED: Stuff-sacks for waterproof gear storage, sleeping bag

CONCESSIONS: Film, personal items and souvenirs can be purchased on the trip.

WHAT NOT TO BRING: Guns, large guitars, thick absorbent foam sleeping pads

YOU WILL ENCOUNTER: Nuchalnuth Village inhabitants, whales, eagles, over 100 species of birds

SPECIAL OPPORTUNITIES: Hiking, archaeological sites and sweat lodge

ECO-INTERPRETATION: Guides interpret intertidal and subtidal life, gray whale feedings and environmental issues.

RESTRICTIONS: Do not take pictures of indigenous peoples without establishing a rapport with them.

REPRESENTATIVE MENU: *Breakfast*—French toast; *Lunch*—Greek salad with pita bread; *Dinner*—fresh salmon, veggie pasta, salad, brownies; *Alcohol*—not actively discouraged or encouraged

ABOUT THE COMPANY

TOFINO EXPEDITIONS
1857 W. 4th Ave., #114, Vancouver, BC V6J 1M4, Canada, (604) 737-2030

Tofino Expeditions, emphasizing sea kayaking trips on the west coast of North America, is a for-profit company owned by Grant Thompson that stresses environmental interpretation and low-impact camping techniques. In business for five years, the outfitter donates to local conservation groups such as Friends of Clayoquot Sound and provides exposure to the Sierra Club in its brochure.

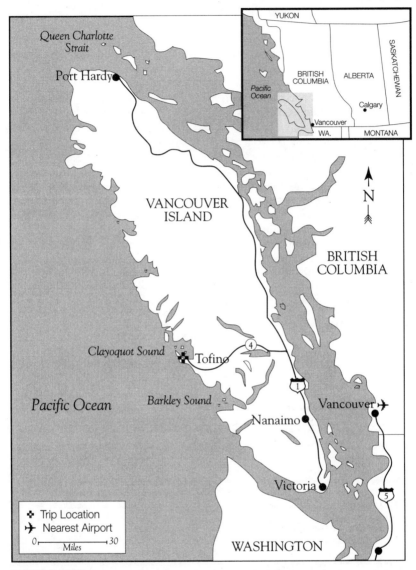

ECOFOCUS The protection of the rainforests from both logging and mining interests is the primary concern in British Columbia as the pace of tree and habitat loss quickens. With the loss of habitat comes the endangering of many of the rainforests' thousands of flora and fauna species.

POLAR BEAR WATCH
View polar bears in Manitoba with Natural Habitat Wildlife Adventures

During summer, the polar bears of Hudson Bay live on the southern shores, scavenging for whatever food they can find. But when autumn comes, the cool air summons the bears northward to their seal-hunting grounds on the ice packs of the bay's northern shore. More than 1,000 bears travel over the tundra where the small grain port of Churchill now stands. In the past, bears were killed by fearful townspeople. Now the influx of tourists has inspired locals to make an enormous effort to protect this valuable population. They have even constructed a polar bear jail to manage problem bears without having to destroy them. Travelers visit the bears in the safety of all-terrain buggies. The bears will approach inquisitively, sometimes resting their paws on the side of the buggy and peeking in at the visitors. Four varieties of the trip are available; the nine-day tour includes a helicopter trip over the preservation area and a visit to the denning site. Guests will stay at comfortable lodgings in Winnipeg and Churchill or at the Churchill Northern Studies Center. The Polar Bear Watch is a priceless opportunity to mingle with these magnificent Arctic bears while playing a part in helping to ensure their continued safety.

ABOUT THE TRIP

LENGTH OF TRIP: 6, 7, 8 or 9 days

DEPARTURE DATES: Oct. 12, 15, 19, 22, 26, 29, Nov. 1

TOTAL COST: $1,745-$2,595

TERMS: Checks; $195 deposit to reserve; cancellation up to 90 days prior to departure

DEPARTURE: Arrival at airport in Winnipeg; air transport to Churchill provided

AGE/FITNESS REQUIREMENTS: All ages; no special physical fitness required

SPECIAL SKILLS NEEDED: Obeying safety rules

TYPE OF WEATHER: 0°-45° (buggies are heated to 45°-50°)

OPTIMAL CLOTHING: Warm boots, warm pants, long underwear, sweater, jacket, mittens or gloves, warm hat

GEAR PROVIDED: All accommodations and transportation

GEAR REQUIRED: Warm clothes, hats, sunglasses, parka, long johns, warm boots

CONCESSIONS: Some film speeds unavailable

WHAT NOT TO BRING: Hard luggage

YOU WILL ENCOUNTER: Extraordinary wildlife, including polar bears, arctic foxes, willow ptarmigans, gyrfalcons, snowy owls and lemmings

SPECIAL OPPORTUNITIES: Sightseeing in Churchill, visits to the Eskimo Museum, the grain elevators, Prince of Wales' Fort, the Arctic Trading Post and the "Polar Bear Jail"; excellent photography

ECO-INTERPRETATION: One full day is devoted to bus touring the Churchill area and visiting the Inuit museum. Guides give slide shows every night on different wildlife areas.

RESTRICTIONS: None

REPRESENTATIVE MENU: *Breakfast*—eggs, different meats, pancakes, coffee; *Lunch*—baked goods, sandwiches, soups, hot chocolate; *Dinner*—vegetarian meals, pasta, steak; *Alcohol*—not provided

ABOUT THE COMPANY
NATURAL HABITAT WILDLIFE ADVENTURES
One Sussex Station, Sussex, NJ 07461, (800) 543-8917
Natural Habitat Wildlife Adventures takes adventurers of all types to see the world's most magnificent animals, wild and free, in their own natural habitat. A for-profit organization in business seven years, it focuses on using ecotourism to replace lost hunting revenue to communities. It concentrates on specific animals in danger and tries to get other companies to join in promoting wildlife tourism. The company also donates to the International Fund for Animal Welfare and the Churchill Northern Studies Center.

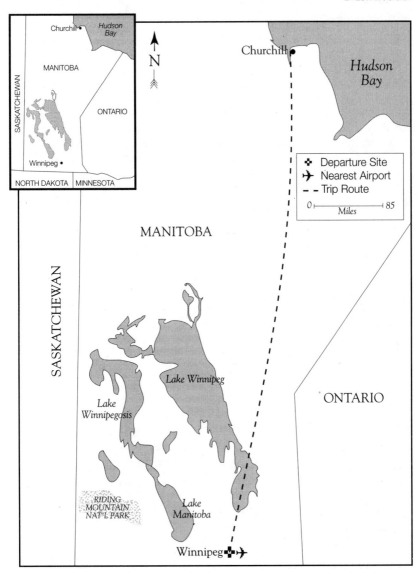

Manitoba

ECOFOCUS Churchill, Manitoba, is known as the "Polar Bear Capital of the World" because there is no other place on Earth where humans can see so many bears with such ease and safety. Fortunately, this increased tourist business has helped change the attitudes of townspeople toward these powerful animals. While this bear population was once in great danger because of its proximity to town, it is now being carefully protected by locals who rely upon the thriving tourist industry.

QUEST FOR NARWHALS
Sea kayak the Northwest Territories with Canada's Canoe Adventures

This is your opportunity to explore the coastline and islands of Admiralty Inlet by sea kayak. Using a fleet of four solo hard-shell and two tandem soft-shell sea kayaks, this trip accommodates ocean explorers of every ability level. A 35-foot fishing boat is used to carry camp staff, food and gear in order to allow you time to enjoy to the fullest extent this terrific northern environment, filled with marine mammals and a plethora of northern birds. Paddling along with a knowledgeable company guide, you will have the opportunity to see incredible wildlife, including polar bears, caribou, Arctic foxes, seals, orca and beluga whales, walruses and, of course, the rare and beautiful narwhal, with its six- to 10-foot unicorn tusk spiraling from the waters as it moves. Also included in this expedition is the extraordinary chance to visit the ancient Dorset and Thule archaeological sites with local Inuit guides and staff from Arctic Bay. Along the fair-wind side of Yeoman Island, you will stop at nesting sites where Arctic tern eggs will be hatching and common eider ducks are in the middle of incubation. On this trip, you will have amazing adventures without sacrificing comfort and leisure.

ABOUT THE TRIP

LENGTH OF TRIP: 8 days

DEPARTURE DATES: Aug. 6, 13

TOTAL COST: $3,665 Canadian

TERMS: Cash, check, VISA, MC; deposit $350 Canadian nonrefundable; reimbursement 75% with cancellation 60 days prior to trip, refund varies according to time of notification

DEPARTURE: Arrive at airport in Nanisivik; further transport by van provided

AGE/FITNESS REQUIREMENTS: 18 years or older; good health

SPECIAL SKILLS NEEDED: None

TYPE OF WEATHER: 50°-70° days; 40°-55° nights

OPTIMAL CLOTHING: Loose, warm clothing, neoprene socks and runners

GEAR PROVIDED: All camping and cooking equipment, kayaks, paddles, personal flotation devices

GEAR REQUIRED: Sleeping bag, clothing, toiletries

CONCESSIONS: Should bring film, but is available first day and upon your return.

WHAT NOT TO BRING: Guns, rifles

YOU WILL ENCOUNTER: Indigenous people (Inuit), incredible wildlife including narwhals, Arctic terns, eider ducks, caribou and Arctic foxes

SPECIAL OPPORTUNITIES: Visiting Dorset and Thule archaeological sites

ECO-INTERPRETATION: Interpretation is provided by local Inuit with an emphasis on human and natural history and narwhal viewing.

RESTRICTIONS: Nanisivik and Arctic Bay are primarily Inuit communities; show cultural appreciation.

REPRESENTATIVE MENU: *Breakfast*—hot breakfasts; *Lunch*—fresh fruit, Inuit dishes, hearty fare of wilderness expeditions; *Dinner*—local delicacies: caribou steaks or Arctic char; *Alcohol*—can be purchased in Arctic Bay or bring your own

ABOUT THE COMPANY

CANADA'S CANOE ADVENTURES—CANADIAN RECREATIONAL CANOEING ASSOCIATION
1029 Hyde Park Rd., Ste. 5, Hyde Park, ONT N0M 1Z0, Canada, (519) 641-1261
Canada's Canoe Adventures is owned and operated by the Canadian Recreational Canoeing Association (CRCA), a nonprofit organization in business 20 years specializing in wilderness tourism, specifically canoe and kayak trips. This company works as a conservation organization with the Canadian government and supports local Inuit co-ops. The CRCA is also responsible for cleanup, environmental scholarship funds, working with the Canadian Heritage River System waterway to designate rivers for protection, community-based work and providing information services to paddlers around the world.

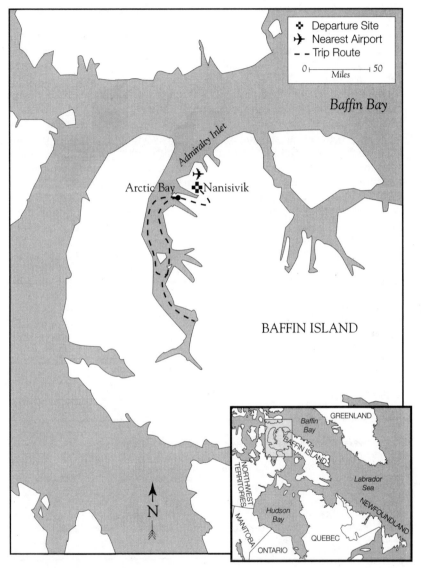

ECOFOCUS The increasing number of people entering this area has an effect on the wildlife. Also, the growing industries of logging and mining, plus acid rain, are damaging the environment.

RAFT THE FIRTH RIVER
Float through the British Mountains with Canadian River Expeditions

If you've ever dreamed of visiting the Arctic, the Firth River Expedition is for you. Located at the northern end of the Yukon Territory between Alaska's Arctic National Wildlife Refuge and the Arctic Ocean, the Firth runs through the British Mountains, emptying into a coastal plain and a broad delta at the Beaufort Sea. You'll float through the interior mountains and coastal plains and see thousands of migrating birds and porcupine caribou. Moose rut in the mountain region while musk-oxen graze on the Arctic plains. You will raft only a few hours a day, so there is plenty of time for hiking on the open tundra. An expert naturalist accompanying the trip will identify species and bring to life the subtle signs of the Inuit hunters who have hunted along the river for millennia. At the delta, you can strap on your hip waders for the crossing to Pauline Cove on Herschel Island, where you'll see a turn-of-the-century whaling station. The wildflowers and birds on Herschel Island are spectacular, and you lay over for a day to watch eiders and guillemots in flight. The Arctic breeze rustles the grasses at the tops of ocean bluffs you'll stand on to watch the ringed and bearded seals play while you scan the waters for beluga and bowhead whales before flying back to Inuvik.

ABOUT THE TRIP

LENGTH OF TRIP: 11 days
DEPARTURE DATES: June 26, July 9, 22
TOTAL COST: $2,900
TERMS: Check; $200 deposit refunded up to 30 days in advance
DEPARTURE: Nearest airport is Inuvik, Northwest Territories. Transportation provided from there.
AGE/FITNESS REQUIREMENTS: 8-80 years old
SPECIAL SKILLS NEEDED: Good health, but no special skills needed
TYPE OF WEATHER: Variable; 24-hour sunlight, 32°-100°
OPTIMAL CLOTHING: Wide range of clothing to reflect variable conditions
GEAR PROVIDED: All except clothing
GEAR REQUIRED: Clothing, camera, binoculars, etc. Bring hip waders for the delta
CONCESSIONS: No concessions can be purchased on the trip.
WHAT NOT TO BRING: Guns or stereo equipment

YOU WILL ENCOUNTER: Ancient Inuit and pre-Inuit cultural sites along the river, fabulous wildlife (up to 100,000 caribou crossing the river)
SPECIAL OPPORTUNITIES: Sleeping on the tundra under the midnight sun; visiting a historic whaling station; innumerable photo opportunities; fishing for arctic grayling and char
ECO-INTERPRETATION: An expert naturalist accompanies each trip. All guides go through an in-house interpretation and natural history training program.
RESTRICTIONS: Trash along the trip will be burned or sorted for recycling and packed out, and care is taken not to spook the animals.
REPRESENTATIVE MENU: *Breakfast*—fresh fruit, pancakes, cereal, sausage; *Lunch*—various cheeses, deli slices, taco salad, sweets; *Dinner*—soup, large fresh salad, beef Stroganoff, fresh-baked chocolate cake; *Alcohol*—wine provided with dinner, beer provided in camp in afternoon

ABOUT THE COMPANY
CANADIAN RIVER EXPEDITIONS
3524 W. 16th Ave., Vancouver, BC V6R 3C1, Canada, (604) 738-4449
A for-profit company in business for 21 years, Canadian River Expeditions is owned and operated by John Mikes Sr. and Johnny Mikes Jr., employing a staff of three to 15. The company specializes in rafting trips in British Columbia, Alaska and the Yukon Territory, using local guides when possible. It was instrumental in launching the campaign to save the Tatshenshini River.

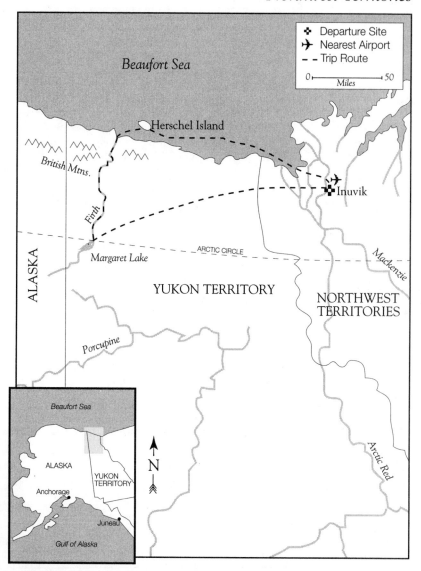

ECOFOCUS Drilling in the Arctic National Wildlife Refuge, and offshore in the Canadian part of the Beaufort Sea, threatens the porcupine caribou herd that fords the Firth River each year on its journey between the wildlife refuge and the Northern Yukon National Park.

SEAL WATCH
Visit endangered harp seals with Natural Habitat Wildlife Adventures

The peaceful, remote Magdalen Islands, located just 75 miles north of Prince Edward Island, supply birthing grounds for nearly 250,000 harp seals. This March, join Natural Habitat Wildlife Adventures and seize this unique opportunity to visit these endearing creatures in their environment. After a brief safety presentation and orientation to the seal habitat, hop aboard a helicopter bound for the ice floes. Fly over this desolate white wilderness and land within walking distance of hundreds of seals. Step out onto the ice and walk among the animals. Some pups are surprisingly friendly and playful and will even roll over and let you pet their bellies. Slide shows and lectures on seals follow, led by Dr. Gary Miller, professor at the University of New Mexico, and photographer/naturalist Steve Morello. Then go for a romp in the snow on snowshoes or cross-country skis. Rejoin the group for dinner and conversation at the local pub. Not only is the Seal Watch a wonderful opportunity to visit some of the world's most beautiful creatures, it is a chance to help those same animals by providing an alternative source of revenue in the hunting communities.

ABOUT THE TRIP

LENGTH OF TRIP: 5-6 days
DEPARTURE DATES: Feb. 29, Mar. 4, 8, 12, 15
TOTAL COST: $1,495-$1,995
TERMS: Checks; $195 deposit; cancellation up to 90 days prior to trip
DEPARTURE: Arrival in Halifax airport; further transport provided
AGE/FITNESS REQUIREMENTS: None
SPECIAL SKILLS NEEDED: None
TYPE OF WEATHER: 0°-40°F
OPTIMAL CLOTHING: Warm clothing
GEAR PROVIDED: One-piece expedition suits and boots
GEAR REQUIRED: None
CONCESSIONS: Available throughout

WHAT NOT TO BRING: Hard luggage
YOU WILL ENCOUNTER: Incredible wildlife viewing, including both harp and hooded seals
SPECIAL OPPORTUNITIES: Fantastic photography; snowshoeing, cross-country skiing (equipment and guides provided at no charge)
ECO-INTERPRETATION: Participants will be provided with daily slide shows and lectures on seals and other wildlife.
RESTRICTIONS: None
REPRESENTATIVE MENU: *Breakfast*—full American-style breakfast; *Lunch*—not included but plenty of restaurants are available; *Dinner*—vegetarian and local seafood specialties; *Alcohol*—no policy; available for purchase

ABOUT THE COMPANY

NATURAL HABITAT WILDLIFE ADVENTURES
One Sussex Station, Sussex, NJ 07461, (800) 543-8917

Natural Habitat Wildlife Adventures takes adventurers of all types to see the world's most magnificent animals, wild and free, in their own natural habitat. A for-profit organization in business seven years, it focuses on using ecotravel to replace lost hunting revenue to communities. It concentrates on specific animals in danger and tries to get other companies to join in promoting wildlife tourism. It also donates to the International Fund for Animal Welfare to promote Seal Watch.

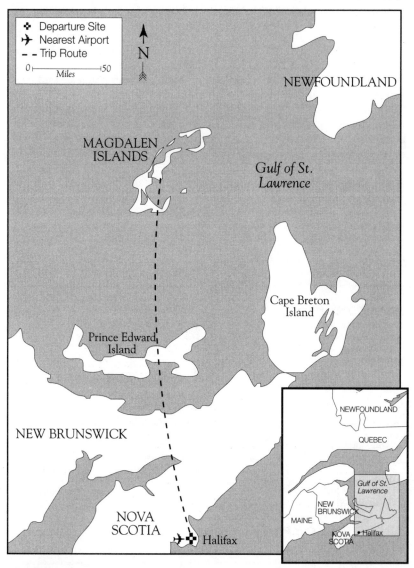

Legend:
- ❖ Departure Site
- ✈ Nearest Airport
- – – Trip Route

0 ———————— 50
Miles

N

NEWFOUNDLAND

MAGDALEN
ISLANDS

*Gulf of St.
Lawrence*

Cape Breton
Island

Prince Edward
Island

NEW BRUNSWICK

NOVA
SCOTIA ✈❖ Halifax

NEWFOUNDLAND

QUEBEC

*Gulf of St.
Lawrence*

NEW
BRUNSWICK
MAINE

NOVA • Halifax
SCOTIA

ECOFOCUS Every February, about 250,000 harp seals journey to eastern Canada's Gulf of St. Lawrence from their Arctic home. This is where they give birth, on ice floes just north of the small fishing villages of the Magdalen Islands. Despite 25 years of public pressure, seal hunting, after a decade-long respite, has been restored in Canada. The harp seal population has increased by 1.5 million in the last seven years and is blamed for the decrease in cod available to the fishing industry.

WHALES AND RESEARCH
Research baleen whales in Quebec with Mingan Island Cetacean Study

Nowhere else will you be able to get this close to a blue whale while actually contributing to valuable research to aid its survival. Research sessions with Mingan Island Cetacean Study offer you the opportunity to get away to the solitude of the sea and immerse yourself in an extraordinary educational adventure. Your home will be the coastal village of Longue Pointe de Mingan, the gateway to the Mingan Islands. These magnificent islands are covered with distinctive limestone rock formations carved by ice and sea. Here you will be introduced to seabirds such as puffins, Arctic terns, murres and eider ducks, as well as to the unique geology and flora of this fascinating archipelago. Early each morning you will join experienced field biologists aboard 24-foot hard-bottom inflatable boats and travel out through the islands. Gray seals and harbor porpoises greet you as you head offshore toward the domain of the great whales. Talks covering the description and biology of marine mammals as well as current research are presented at the field station as part of research sessions. Experiencing the graceful breach of a 40-ton humpback whale, a close encounter with an 85-foot blue whale or being engulfed in a herd of 300 white-sided dolphins—how could this not be the adventure of a lifetime?

ABOUT THE TRIP

LENGTH OF TRIP: 7-21 days
DEPARTURE DATES: June 15-Oct. 25
TOTAL COST: $139 per day
TERMS: Cash, check; 40% deposit; cancellation fee depends on amount of advance notice
DEPARTURE: Transport from airport in Sept-Iles provided in company van
AGE/FITNESS REQUIREMENTS: 13-75 years of age; good health, though special medical needs can be met
SPECIAL SKILLS NEEDED: None
TYPE OF WEATHER: Expect wind, sun, rain, fog; 50°-75°
OPTIMAL CLOTHING: Slickers, layered clothing, hat, sunglasses and gloves
GEAR PROVIDED: Flotation suits or vests

GEAR REQUIRED: None
CONCESSIONS: Souvenirs available in Longue Pointe de Mingan; film should be purchased prior to departure
WHAT NOT TO BRING: Business clothing
YOU WILL ENCOUNTER: Indigenous people and plenty of wildlife
SPECIAL OPPORTUNITIES: Fishing
ECO-INTERPRETATION: All participants are accompanied by researchers with a minimum of two years experience.
RESTRICTIONS: None
REPRESENTATIVE MENU: *Breakfast*—cereal, eggs, juice; *Lunch*—sandwiches; *Dinner*—restaurants with a wide selection; *Alcohol*—available for purchase; not allowed at sea

ABOUT THE COMPANY

MINGAN ISLAND CETACEAN STUDY
285 Green, St. Lambert, QC J4P 1T3, Canada, (514) 465-9176
Mingan Island Cetacean Study is a US and Canadian nonprofit research organization in business for 13 years and involved in marine mammal studies, education and conservation. All funds raised go back into the operation of the station. Visitors get to participate in the use of their financial contribution and work directly with biologists.

Quebec

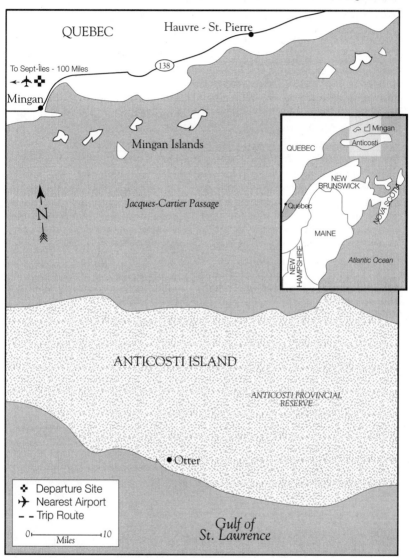

ECOFOCUS There is a great deal of pollution in the busy St. Lawrence Seaway, causing toxins in whale blubber. Blubber biopsies show how far pollution spreads and how it is affecting animals. Also, ships in the area have struck whales, scarring 20 percent of them. In addition, fishing gear entangles animals.

BAHAMAS—THE COMPLETE EXUMAS
Sea kayak in the Bahamas with Ecosummer Expeditions

During this excursion, sea kayakers will paddle through crystal-clear water, by milky, white, sandy beaches, and camp in some of the 350 small, mostly uninhabited cays, as they cover over 100 miles of this spectacular chain of islands. While neighboring islands experience heavy tourism, some of the Exumas cays are protected as the Exuma Cays Land and Sea Park and under the auspices of the Bahamas National Trust. Ecosummer Expeditions donates to the trust and gives allowances to the park to ensure proper management. During the trip, paddlers will average about 10 miles a day, usually in the morning, before the prevailing wind kicks in. That allows trip participants to beach their kayaks and either hike through the many trails to discover the distinct flora and avian fauna of the islands, or strap on snorkeling gear to explore the wide variety of marine life. The snorkeling is especially superb—in some areas there will be massive star, brain and elkhorn corals, while others contain sea caves, mangrove-lined tidal creeks or wave-swept rocky reefs covered with swaying sea fans. Also tempting is the mysterious allure of underwater cave diving.

ABOUT THE TRIP
LENGTH OF TRIP: 14 days

DEPARTURE DATES: Feb.-Mar.

TOTAL COST: $1,995

TERMS: All types of payment accepted; $400 deposit 60 days prior; full refund less $100 fee if cancelled 60 days prior to trip

DEPARTURE: Arrive at George Town airport; taxi to hotel 10 miles from airport

AGE/FITNESS REQUIREMENTS: Reasonably fit; no major medical problems

SPECIAL SKILLS NEEDED: None

TYPE OF WEATHER: Temperatures of 75°; less than 2″ monthly rainfall

OPTIMAL CLOTHING: Running shoes, amphibious shoes, lightweight clothing and raingear

GEAR PROVIDED: Tents, paddles, kayaks and life jackets

GEAR REQUIRED: Personal gear, clothing and sleeping bag

CONCESSIONS: Film should be purchased prior to departure; personal items and souvenirs can be bought in George Town and Nassau.

WHAT NOT TO BRING: Formal attire, hard suitcases

YOU WILL ENCOUNTER: Many species of birds

SPECIAL OPPORTUNITIES: Snorkeling and cave diving

ECO-INTERPRETATION: None

RESTRICTIONS: None

REPRESENTATIVE MENU: *Breakfast*—oatmeal, granola, pancakes, eggs and bacon; *Lunch*—assorted sandwiches; *Dinner*—burritos, vegetarian chili, spaghetti and curried rice; *Alcohol*—some provided

ABOUT THE COMPANY
ECOSUMMER EXPEDITIONS
1516 Duranleau St., Vancouver, BC V6H 3S4, Canada, (604) 669-7741; (800) 688-8605
Specializing in sea kayaking, rafting, canoeing and trekking to northern North America and in tropical islands, Ecosummer is dedicated to keeping the adventure in adventure travel. This 16-year-old company, owned by Jim Allan, donates membership dues to Exumas Cays Land and Sea Park and to Bahamas National Trust in the name of its guests.

The Bahamas

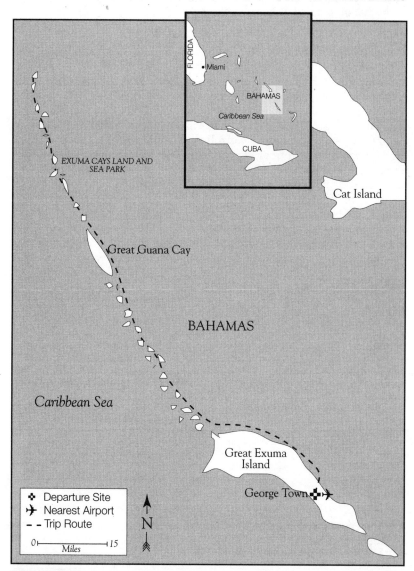

ECOFOCUS While the Exuma Cays Land and Sea Park has been granted protected status and watched over carefully by the Bahamas National Trust, its environmental integrity has still being threatened. The danger stems not from overuse of the islands' land areas, but from overfishing of its marine life.

JAMAICAN EXPLORATION
Birdwatch in Jamaica with Cal Nature Tours

From a villa built by Alex Haley, the author of *Roots* and *Malcolm X*, overlooking Savanna La Mar and tucked 25 miles into the Jamaican forest, you will begin the expedition. The trip includes three excursions, including the Black River Safari trip, an exploration of coastal Negril and a visit to some unique Jamaican limestone caves. At Negril you will have the opportunity to examine the tidal marine life and shop for local crafts. If the ocean intrigues you, don a mask and swim among the intricate coral reefs just off the coast. The next day, stroll through the vast gardens and enjoy the exotic bird life and tropical flora. There are 160 bird species in the surrounding forests and over 20 types of wild fruit. Your Jamaican guide will take you to local farms and explain the local farming economy and the necessity of maintaining the hardwood forest ecosystem. Set off to the Black River National Park for a safari boat ride. The guides will show you over 20 species of birds, as well as feeding crocodiles living in this river and marsh area. The journey would not be complete without visiting scenic waterfalls and a cave reported to have healing mineral waters.

ABOUT THE TRIP

LENGTH OF TRIP: 7 days

DEPARTURE DATES: Year-round

TOTAL COST: $980 with airfare from Miami or Atlanta

TERMS: Cash; $200 deposit; cancellation fee of $50 if notification is given at least 60 days in advance

DEPARTURE: Arrive in Montego Bay and travel 25 miles into the mountains by mini-bus to Villa Tambrin Hill.

AGE/FITNESS REQUIREMENTS: General good health

SPECIAL SKILLS NEEDED: None

TYPE OF WEATHER: Warm, occasional light rain; 80°-85°

OPTIMAL CLOTHING: Light, casual clothes

GEAR PROVIDED: None

GEAR REQUIRED: Binoculars, insect repellent, sun block, hat

CONCESSIONS: Should be purchased before you depart

WHAT NOT TO BRING: Fancy clothing, excess baggage

YOU WILL ENCOUNTER: Numerous species of birds, crocodiles, fruit bats, incredible flowers, limestone caves

SPECIAL OPPORTUNITIES: Fishing, swimming and photography

ECO-INTERPRETATION: Discussion of deforestation, wetland loss and ocean pollution

RESTRICTIONS: None

REPRESENTATIVE MENU: *Breakfast*—tropical fruit and Jamaican bread; *Lunch*—fruit, bread; *Dinner*—native food: rice, chicken, seafood, fruit, rum ice cream; *Alcohol*—Jamaican rum and beer are available

ABOUT THE COMPANY

CAL NATURE TOURS, INC.
7310 S.V.L. Box, Victorville, CA 92392, (619) 241-2322
Cal Nature Tours is a for-profit company, owned by Thomas Gwin, in operation for seven years with a staff of seven. Contributions to the preservation of Jamaica are made through donations to the Black River wetlands, to the foundation of Black River National Park and to several US environmental groups. Five percent of the trip proceeds are donated to Jamaican conservation groups and to local guides. Cal Nature Tours is also involved in a reforestation project.

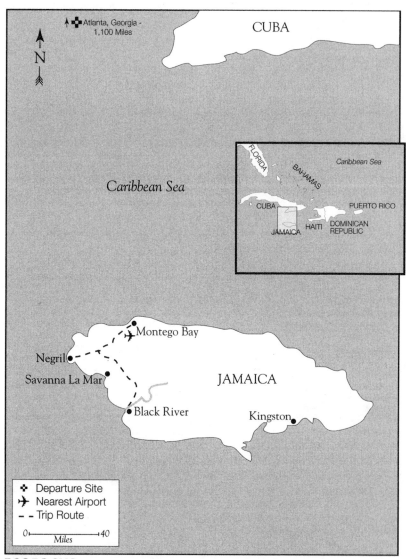

ECOFOCUS Deforestation threatens Jamaica's intricate environment. The depletion of coral reefs due to the impact of irresponsible scuba divers is also a threat to this delicate ecosystem.

CARIBBEAN RAINFOREST HIKING TOUR
Ecological hiking in Dominica with Earth Tours

Hosted by the Earth Tours staff and several local naturalist guides, Caribbean Rainforest Hiking will take you along the path of the Carib Indians on a discovery of Dominica. This pristine, unspoiled Caribbean island is located between Guadalupe to the north and Martinique to the south. Pastoral and exotic, this country contains some of the most stunning scenery imaginable. Beaches as black as coal lie side by side with white sand beaches the color of pearls. Majestic mountains tower over lush, verdant valleys. Mighty rivers transform from meandering streams into thundering waterfalls. All of these are yours to explore on this Earth Tours expedition. Daily activities will range from moderate hiking to a leisurely visit to catch a glimpse of the endangered Sisserou parrot–only 80 to 100 remain in Dominica and the rest of the world. There will be plenty of time to swim in crystal waterfalls, lakes and rivers and, of course, to snorkel in the Caribbean itself. Half of your time will be spent in accommodations on the island interior, surrounded by rainforest, and the other half right on the beach. All ground services and meals are provided, with the exception of one lunch on the day you visit the capital of Roseau.

ABOUT THE TRIP

LENGTH OF TRIP: 7 days

DEPARTURE DATES: Monthly year-round (winter months have multiple trips)

TOTAL COST: $990

TERMS: Cash and credit cards accepted for a $300 deposit and the remainder of the fee

DEPARTURE: Tour departs from Dominica's only airport; all transportation provided by Earth Tours

AGE/FITNESS REQUIREMENTS: None

SPECIAL SKILLS NEEDED: Good general health and hiking ability

TYPE OF WEATHER: Warm and sunny with brief showers in the rainforest

OPTIMAL CLOTHING: Sturdy hiking boots, sneakers, river sandals or Aquasox

GEAR PROVIDED: None

GEAR REQUIRED: Proper footwear, raingear, swimwear, sunscreen, insect repellent

CONCESSIONS: Film available, but it is recommended that travelers bring their own

WHAT NOT TO BRING: Heavy, warm or dress clothing

YOU WILL ENCOUNTER: Constant contact with indigenous people; viewing of magnificent bird life, including the endangered Sisserou parrot

SPECIAL OPPORTUNITIES: Snorkeling, birdwatching, visiting Boiling Lake

ECO-INTERPRETATION: Lecture given by naturalist guide, historian, author and native Dominican on the first night of the tour

RESTRICTIONS: Should dress in proper attire when away from the beach

REPRESENTATIVE MENU: *Breakfast*—English/Caribbean full breakfast; *Lunch*—local fish and fruits; *Dinner*—local seafood delicacies; *Alcohol*—available everywhere except while hiking

ABOUT THE COMPANY

EARTH TOURS LTD.
93 Bedford St., Box 3C, New York, NY 10014, (212) 675-6515
During its two years of existence, Earth Tours Ltd. has promoted ecological rainforest hiking in the Caribbean. This for-profit company, owned by Kevin and Marsha Vaughn, is a member of the Dominica Conservation Association, to which it contributes financially and administratively. Earth Tours Ltd. promotes local businesses and attempts to put most of its profits back into the local economy.

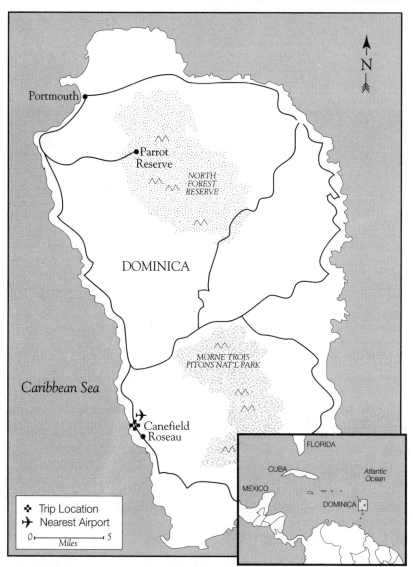

ECOFOCUS Dominica and other unspoiled Caribbean areas are endangered by the trappings of traditional tourism–high-rise condos and golf courses. Work is being done to limit commercialization of these islands, in order to protect the land and the natural habitat of local endangered species.

GRASSROOTS CARIBBEAN
Sail, hike and snorkel in the Caribbean with Interlocken

Plunge into the blue-green waters of the Caribbean Sea on a marine adventure, exploring the delicate balance between culture and environment. From San Juan, you will venture to the remote island of Culebra, where you will meet the local islanders, learn about the plight of the giant sea tortoise and practice your Spanish with elementary school children at a summer day-camp. After Culebra, you will travel to the Virgin Islands, where you will have the opportunity to explore the arts, crafts and music of the indigenous culture. The island of St. John is almost entirely a wilderness preserve, under the protection of the National Park Service. There you will camp and learn about the balance maintained between human beings and nature. After visiting St. Thomas and the British islands of Tortola and Virgin Gorda, you will set sail on a 51-foot sailboat for the islands of the eastern Caribbean, including Anguilla, St. Martin and St. Barts. The boat will anchor in coves, allowing you to venture ashore in a small skiff to explore the villages and wildlife of some of the most unspoiled islands in the region. You will have ample opportunity to swim and snorkel. The final stop will be the island of Anguilla, where a community service project will complete your four-week Caribbean exploration.

ABOUT THE TRIP

LENGTH OF TRIP: 28 days

DEPARTURE DATE: June 30

TOTAL COST: $2,975

TERMS: Cash, check; deposit of $500; cancellation policy varies according to time of notification, but after June 15 no refund is guaranteed.

DEPARTURE: Arrive in San Juan, Puerto Rico, and meet group at airport to travel to Culebra Island

AGE/FITNESS REQUIREMENTS: Open to students in grades 9 through 12; good physical condition

SPECIAL SKILLS NEEDED: Spanish and swimming skills are helpful.

TYPE OF WEATHER: Warm, humid tropical climate; 75°-85°

OPTIMAL CLOTHING: Lightweight clothing, swimsuit

GEAR PROVIDED: Technical gear is provided.

GEAR REQUIRED: Lightweight sleeping bag, day pack, backpack, Spanish dictionary, walking/hiking shoes, snorkeling gear

CONCESSIONS: Available throughout the trip

WHAT NOT TO BRING: Walkman

YOU WILL ENCOUNTER: Giant sea tortoises, plentiful marine and animal life

SPECIAL OPPORTUNITIES: Family stay, community service in Anguilla, counseling at day camp for underprivileged children

ECO-INTERPRETATION: Discussion led by an environmental-education specialist about the delicate balance that exists between people and the Caribbean environment.

RESTRICTIONS: Environmental and cultural restrictions will be discussed.

MEALS: Indigenous Caribbean cuisine; group selects several meals at local markets

ABOUT THE COMPANY

INTERLOCKEN
RR 2, Box 165, Hillsboro, NH 03244, (603) 478-3166
Interlocken is a for-profit company with two nonprofit branches, The Global Roots Community Service and The Educational Opportunities Fund. Interlocken is run by Richard Herman, the founding director, and has been in business for 32 years with a staff of 150 people. Specializing in community-service/educational/adventure tours to the US, Asia, Africa and Latin America, Interlocken contributes to environmental preservation through international community-service projects.

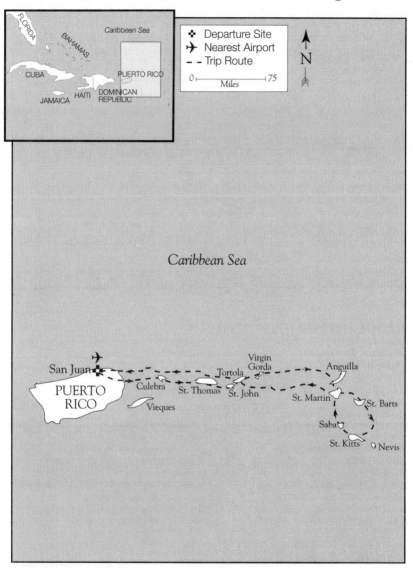

ECOFOCUS Coral reefs in the Caribbean are threatened by pollution, tourism and global warming. The low-nitrate environment of a healthy reef is offset by sewage dumped in the ocean. Tourism adversely affects the reefs because of the huge number of people who anchor boats, remove shells and coral, and spear fish. Global warming is decreasing the reef's growth levels and tolerance to disturbance caused by human beings and marine life. Increased water temperatures and increased production of oxygen are also killing the reefs.

THE MONKEY BAY WILDLIFE SANCTUARY
Sea kayak, snorkel, raft, hike and ride horseback in Belize with R.A.I.N.

Explore the ecosystems of Belize while indulging in a variety of exciting activities, including sea kayaking or sailing, swimming, river rafting, horseback riding, birdwatching and nature walking. You will have the opportunity to explore Belize's pristine tropical environment managed as parks, sanctuaries and forest reserves, for which 38 percent of the land is allotted. Monkey Bay Sanctuary, where you will camp and hike, is a non-governmental reserve and a role model for private land stewardship. The sanctuary covers 1,070 acres of tropical forest and savanna, bordered by the Sibun River, which flows from the Maya Mountains through the coastal savanna en route to the Caribbean Sea. Outside the sanctuary, you can sea kayak or sail and snorkel through the largest barrier reef in the western hemisphere, and raft through virgin jungle and river caves. If your need for adventure has not been satiated you can visit the ancient Mayan ruins and temples or hike through the Cockscomb Basin Wildlife Sanctuary and Jaguar Preserve. Culturally oriented travelers will enjoy a visit to the Ix Chel Farm and Panti Medicinal Trail, where the vast botanical and medicinal knowledge of the Maya are being documented and preserved. You will also have a chance to examine the jungle wildlife and climb the pine ridges by horseback.

ABOUT THE TRIP

LENGTH OF TRIP: 10 days

DEPARTURE DATES: Feb.-Apr.

TOTAL COST: $1,400

TERMS: Check; deposit of $200; cancellation policy requires 3 weeks advance notice for refund

DEPARTURE: Arrive in Belize City at the Phillip Goldson International Airport and meet group vans.

AGE/FITNESS REQUIREMENTS: Good physical condition

SPECIAL SKILLS NEEDED: None

TYPE OF WEATHER: Tropical climate during dry season

OPTIMAL CLOTHING: Light and rugged

GEAR PROVIDED: Sea kayaks, rafts

GEAR REQUIRED: Tent, backpack, water bottle, camping accessories, insect repellent

CONCESSIONS: Available throughout the trip

WHAT NOT TO BRING: Food

YOU WILL ENCOUNTER: Local Maya Indians and Creoles; incredible marine and land life

SPECIAL OPPORTUNITIES: Nature walks, birdwatching, photography, rafting, horseback riding and kayaking or sailing

ECO-INTERPRETATION: The guides will provide information on the native wildlife and flora as well as the Belizean culture and natural history.

RESTRICTIONS: Respect and consideration of local customs

REPRESENTATIVE MENU: *Breakfast*—fruit, cereal; *Lunch*—grains, vegetables; *Dinner*—seafood, vegetables, grains and other wholesome cooking; *Alcohol*—not available

ABOUT THE COMPANY

R.A.I.N.
P.O. Box 4418, Seattle, WA 98104, (206) 324-7163, (206) 633-3913

R.A.I.N. is a nonprofit venture that has provided nature tours for two years with a staff of six people. Specializing in low-impact trips in Belize that support the local economy, help to fund the Monkey Bay Wildlife Sanctuary and educate travelers about the local ecology, R.A.I.N. uses local guides and accommodations.

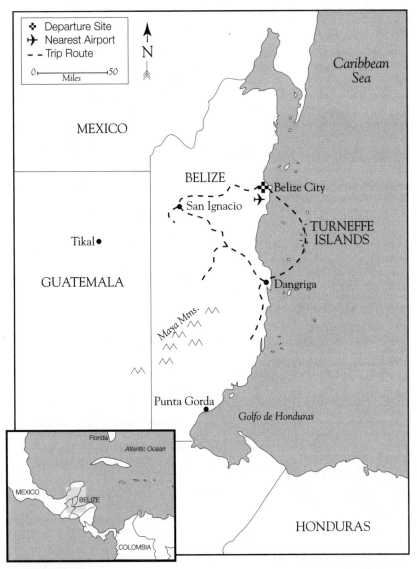

ECOFOCUS The adverse effects of fertilizers, pesticides and sediment washing into the sea are compounded by the effects of overfishing and the increasing numbers of scuba divers and snorkelers.

RAFTING THE PACUARE RIVER
Raft and hike in Costa Rica with Rios Tropicales

The Pacuare River pools and drops through the Costa Rican rainforest like liquid jade. This pristine river has virtually no road access; gear is loaded onto horses belonging to a local farmer and everyone walks the last couple of miles to the put-in. Then the fun begins: Thunderous Class III-IV rapids alternate with lazy stretches of water where you can watch the electric-blue morpho butterflies flit along the emerald banks. The next day, you can hike to visit Cabecar indigenous people living nearby, as well as to view a reforestation project on land owned by Rios Tropicales and worked by local farmers. The third and final day takes you through the Huacas, the two biggest rapids of the trip, both Class IV. There is a beautiful walled pool with a waterfall threading delicately down to the river between them. Then it's on to Dos Montañas Gorge, which is the planned site for a hydropower project that will inundate everything you have seen in the last three days. Rios management and staff have been among the leaders of the opposition to the dam, which is slated for construction within a decade. Opponents are calling for a conservation-minded national energy policy; they also organized a protest at the dam site involving everyone from Boy Scouts to the Cabecars.

ABOUT THE TRIP

LENGTH OF TRIP: 2-3 days

DEPARTURE DATES: All year

TOTAL COST: $220-$300

TERMS: Cash, check, major credit cards; 30% deposit; refund available if cancellation more than 15 days prior

DEPARTURE: Arrival at airport in San José; further transportation can be arranged through company office

AGE/FITNESS REQUIREMENTS: Must be at least 12 years old

SPECIAL SKILLS NEEDED: None

TYPE OF WEATHER: Dry and low water Jan. to Apr., rain May to Dec.

OPTIMAL CLOTHING: Tennis shoes, light rafting clothing

GEAR PROVIDED: Tents, pads, eating and cooking utensils

GEAR REQUIRED: Personal clothes, camping flashlight, sheet or light sleeping bag

CONCESSIONS: Personal items, film and souvenirs may be purchased in San José before the trip begins.

WHAT NOT TO BRING: Wet suit

YOU WILL ENCOUNTER: Indigenous Cabecars; incredible wildlife

SPECIAL OPPORTUNITIES: Hiking through lowland tropical forests, wonderful side creeks

ECO-INTERPRETATION: Guides have knowledge of most bird, mammal, reptile and insect life and will discuss a reforestation program.

RESTRICTIONS: None

REPRESENTATIVE MENU: *Breakfast*—tropical fruit and toast; *Lunch*—taco salad, river buffet; *Dinner*—famous jungle chicken; *Alcohol*—jungle punch provided; bring other if desired

ABOUT THE COMPANY

RIOS TROPICALES
P.O. Box 472-1200, San José, Costa Rica, (506) 33-6455

Rios Tropicales, a for-profit company in business seven years, was created by Rafael Gallo, Fernando Esquivel and James Nixon with the objective to share whitewater with Costa Ricans and the rest of the world. Since then, it has become the largest whitewater rafting and sea kayaking outfitter in Latin America. They use local guides and are prime organizers of the movement against damming the Pacuare River.

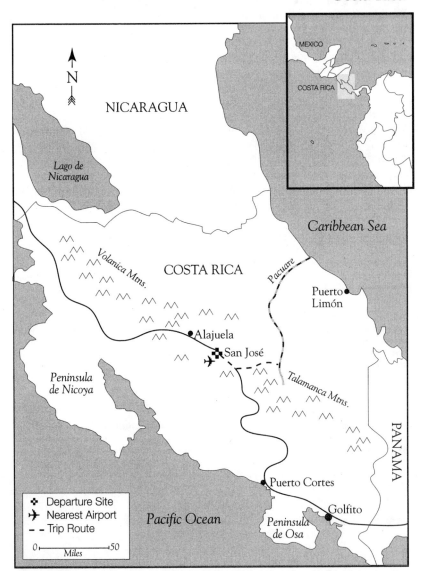

Costa Rica

N

NICARAGUA

Lago de
Nicaragua

Caribbean Sea

Volanica Mtns.

COSTA RICA

Pacuare

Puerto
Limón

Alajuela

San José

Talamanca Mtns.

Peninsula
de Nicoya

PANAMA

Puerto Cortes

Golfito

✤ Departure Site
✈ Nearest Airport
- - Trip Route

Pacific Ocean

Peninsula
de Osa

0 ⊢————⊣ 50
Miles

ECOFOCUS A hydro dam project is being planned in this area. This would destroy a rich ecological zone and relocate its inhabitants. Costa Rica's rivers are also threatened by pollution and deforestation. The Pacuare and Reventazon rivers, home of Costa Rica's whitewater tourism, are threatened by four hydroelectric dams slated to be built over the next ten years.

RARA AVIS RAINFOREST RESERVE
Walk the Costa Rican rainforest with Rara Avis

Rara Avis, a jungle preserve in northern Costa Rica, is testament to the belief that a standing rainforest can be financially productive. Dedicated to tourism, scientific research and the extraction of forest products such as wicker and ornamental plants, Rara Avis has become a significant source of income for nearby residents. Started by North American biologist Amos Bien, the preserve has extensive trails and a beautiful three-tiered waterfall. During your visit to this virgin rainforest, you will have the opportunity to swim under these spectacular waterfalls and enjoy nature walks with knowledgeable naturalists, experiencing some of the most extraordinary wildlife on earth. Rara Avis is home to some 335 species of birds, monkeys, coatis, anteaters, tapirs, peccaries and even the elusive jaguar. It also boasts a contraption straight out of *Alice in Wonderland*: a suspended metal cage called The Automated Web for Canopy Exploration, which lets you travel 150 feet above the forest floor to inspect the jungle canopy. Rara Avis lies a grueling nine miles from the trailhead; the lodge provides transportation via tractor or horseback. A visit to this spectacular tropical haven will provide you with insight on rainforest ecology while helping to preserve these valuable natural resources.

ABOUT THE TRIP

LENGTH OF TRIP: 2 nights minimum
DEPARTURE DATES: All year
TOTAL COST: Approximately $70/night
TERMS: Credit cards, cash, check, traveler's check; 30% deposit; 100% refund with 30 days' notice
DEPARTURE: Nearest airport is San José. Taxis are available in early mornings to Horquetas de Sarapiqui.
AGE/FITNESS REQUIREMENTS: Any age, but quite fit
SPECIAL SKILLS NEEDED: Mud walking
TYPE OF WEATHER: Rain
OPTIMAL CLOTHING: Rubber boots (essential), clothes to get dirty and wet in, light raingear
GEAR PROVIDED: Rubber boots up to size 11 men's (U.S.)
GEAR REQUIRED: Flashlight, changes of clothes, footwear easy to put on and remove

CONCESSIONS: Film and personal items are available in San José.
WHAT NOT TO BRING: Electrical equipment, heavy baggage, fragile items
YOU WILL ENCOUNTER: Birds, monkeys, anteaters and tracks of much more wildlife
SPECIAL OPPORTUNITIES: Swimming at the base of waterfalls, riding horses
ECO-INTERPRETATION: All visitors are provided with an in-house naturalist guide who leads intensive nature walks every day.
RESTRICTIONS: No collecting of plants or animals is allowed.
MEALS: All meals served family-style at one long table

ABOUT THE COMPANY

RARA AVIS S.A. - RAINFOREST LODGE AND RESERVE
P.O. Box 8105-1000, San José, Costa Rica, (506) 53-0844
Rara Avis, a for-profit company in operation for nine years and employing a staff of 15 to 18, specializes in guided nature walks in its private rainforest reserve in Costa Rica. The president of Rara Avis, Amos Bien, has lobbied with banks, governments, travel companies and non-governmental organizations to make rainforests outside the national parks economically productive while remaining intact. Eighty percent of company revenues stay in the region, and the company buys nearly all its goods and services from a nearby small town. Long-term employees also become stockholders.

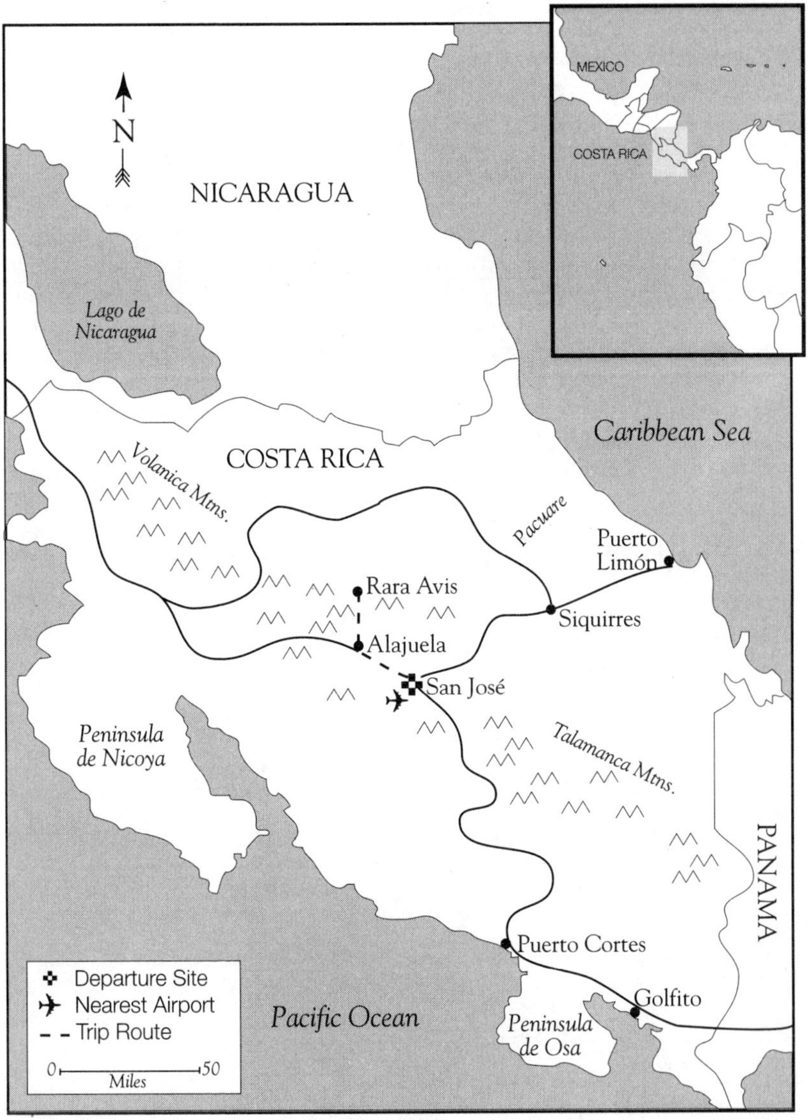
MEXICO
COSTA RICA
N
NICARAGUA
Lago de Nicaragua
COSTA RICA
Volanica Mtns.
Caribbean Sea
Pacuare
Puerto Limón
Rara Avis
Siquirres
Alajuela
San José
Talamanca Mtns.
Peninsula de Nicoya
PANAMA
Puerto Cortes
Golfito
Departure Site
Nearest Airport
Trip Route
0
50
Miles
Pacific Ocean
Peninsula de Osa

ECOFOCUS The area surrounding this reserve is being rapidly deforested. Much of this destruction of rainforest land occurs for the development of low-yield subsistence cattle ranches which yield as little as $7 per acre per year. Rara Avis has problems of trail and road erosion and runoff which are being attended to.

THE SANTA ELENA BIOLOGICAL PRESERVE
Explore Costa Rica's Santa Elena Rainforest with Earth Ventures

Spectacular views of brilliant flora and fauna found nowhere else in the world bring adventure travelers from all corners of the globe to Costa Rica's magnificent cloud forest. Thanks to environmentally responsible planning and coordination of the local community of Santa Elena, the Santa Elena Community-Operated Rainforest Preserve opened in spring of 1992, providing a breathtaking model of ecotourism. This unique community involvement–centered at the local high school–educates people about nature while investing in preservation. Your exploration of this tropical wonderland can be through guided tour or on your own, by horse or bike. The preserve is home to rare, colorful birds, monkeys, sloths, orchids and other flowers. Accommodations can be made at guest houses in the Santa Elena pueblo. You will also have the opportunity to visit the pueblo of nearby Monteverde or see the Monteverde Cheese Factory. On your way back to San José, stop by local craft displays and enjoy beautiful native artistry. Many variations on this trip are available; tour operators can customize trips to accommodate specific desires.

ABOUT THE TRIP

LENGTH OF TRIP: 3 days

DEPARTURE DATES: Year-round beginning spring 1992

TOTAL COST: $475; varies with tour operator

TERMS: Reserve with a $150 check; no cancellation fee with at least 30 days' notice

DEPARTURE: From the Juan Santamaria airport in San José or hotel

AGE/FITNESS REQUIREMENTS: None

SPECIAL SKILLS NEEDED: Healthy and adventurous

TYPE OF WEATHER: Dry from Dec. to April; rainy from May to Nov.; temperate from Dec. to Apr.; average temperature ranges from 50° to 70° in San José

OPTIMAL CLOTHING: Light clothes and sweater with a hat and raingear

GEAR PROVIDED: Boots, raingear, bicycles, horses available to rent

GEAR REQUIRED: Extra shoes

CONCESSIONS: Film may be purchased anywhere, lots of local crafts, cultural promotion, tourism

WHAT NOT TO BRING: More than one bag

YOU WILL ENCOUNTER: Quaker settlers, 336 types of birds, 1,500 kinds of orchids and ferns, sloths, monkeys and reptiles

SPECIAL OPPORTUNITIES: Horseback riding, bicycling and scientific research

ECO-INTERPRETATION: Various levels of explanation are provided, from basic to technical environmental and biological instruction.

RESTRICTIONS: Show respect to locals.

REPRESENTATIVE MENU: *Breakfast*—juice, eggs, fruit and coffee; *Lunch*—rice, beans, vegetables, fruit; *Dinner*—soup, salad, seafood, rice, beans, meat; *Alcohol*—available in town but not on the preserve

ABOUT THE COMPANY

EARTH VENTURES
3700 Trenton Rd., Raleigh, NC 27607, (919) 833-6067
Earth Ventures is a new travel consultant that is dedicated to wilderness travel, nature appreciation, cultural awareness and local support. It researches adventure-tour operators of all types in search of true ecotourism, using specific travel ethics. Earth Ventures donates a percentage of its profits to environmental and cultural groups. Tours make use of resources such as local guides and locally owned homes and lodges. Julio Guzman serves as ecotourism consultant. He is a native Costa Rican with a master's degree in environmental management.

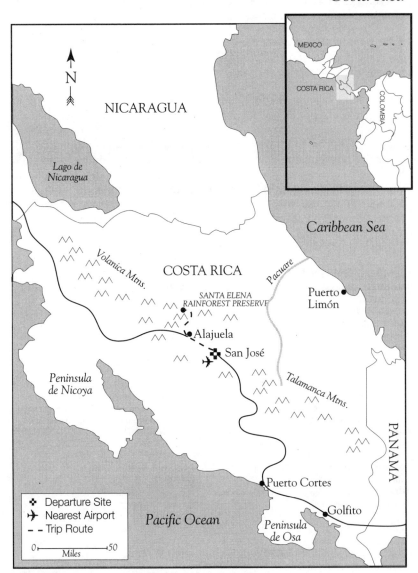

Costa Rica

Legend:
- ✛ Departure Site
- ✈ Nearest Airport
- - - Trip Route

0 |——————| 50
Miles

ECOFOCUS This region suffers from deforestation and negative impacts of increased tourism. The Santa Elena Biological Preserve Rainforest Project strives to decrease forest degradation by responsibly handling the flow of visitors to its national preserve.

WILDLANDS CAMPING SAFARI
Camp and hike through Costa Rica with Wildland Adventures

Exploration of the Costa Rican wildlands will take you through some of this country's richest regions. Beginning in the dry, tropical forests of Guanacaste and moving through Palo Verde National Park, this safari will lead you through country inhabited by camouflaged iguanas, colorful bird life, multicolored frogs and white-faced monkeys. Familiarize yourself with the sights and sounds of nature during your visit to Los Tigres Private Reserve. Arenal Volcano may surprise you with one of its loud rumbles as you are hiking around it or soothing your muscles in a cascading waterfall or natural hot spring. Snorkeling along the beaches of Gandoca-Manzanillo Wildlife Refuge offers a look at varied sea creatures. Walk through the Cocles Indian Reserve, guided by a Bribri Indian naturalist who can point out and explain the local flora and fauna. After a day in the forest, you will appreciate a meal of local flavor prepared and served to you in the home of a native. Camping under the stars and listening to the nocturnal wildlife is a perfect end to a day on safari.

ABOUT THE TRIP

LENGTH OF TRIP: 13 days

DEPARTURE DATES: Feb. 15 and Mar. 28, 1993

TOTAL COST: $1,545

TERMS: Check, MC, VISA, AmEx for $300 deposit; check only for the rest of the land costs; full refund less $100 if canceled 60 days or more prior to departure

DEPARTURE: San José, Costa Rica

AGE/FITNESS REQUIREMENTS: None

SPECIAL SKILLS NEEDED: Previous camping experience a plus but not necessary

TYPE OF WEATHER: 62°-85°, dry and tropical, rain possible

OPTIMAL CLOTHING: Cool clothing, footwear with good traction and ankle support

GEAR PROVIDED: Tents, sleeping bags, pads, kitchen utensils

GEAR REQUIRED: Personal effects including insect repellent and sun protection

CONCESSIONS: Film, personal items and souvenirs can be bought in San José.

WHAT NOT TO BRING: Heavy items or excessive baggage

YOU WILL ENCOUNTER: Indigenous peoples and their culture and an array of wildlife including tropical birds, sloths, iguanas, white-faced monkeys, deer and coatis mundis

SPECIAL OPPORTUNITIES: Dining in local homes, rainforest hikes by Bribri Indian guides, nocturnal rainforest walks and hot springs

ECO-INTERPRETATION: Cultural information is provided by the local guides, while the naturalist guide gives natural history background, wildlife descriptions and explanations of the complex interactions of flora and fauna in a tropical ecosystem.

RESTRICTIONS: Show respect for the locals, who tend to be open and friendly.

REPRESENTATIVE MENU: *Breakfast*—eggs, cereal, juice; *Lunch*—sandwich, fruit, drinks, sweets; *Dinner*—pasta, chicken, rice, beans, tortillas; *Alcohol*—available along the way, but not included

ABOUT THE COMPANY

WILDLAND ADVENTURES
3516 NE 155th St., Seattle, WA 98155, (206) 365-0686, (800) 345- 4453
Wildland Adventures offers authentic worldwide nature and cultural expeditions to focus the efforts of travelers who wish to aid global conservation and support indigenous peoples. For the last 12 years this for-profit company, owned by Kurt Kutay, has led ecotours through the Americas, Africa, Asia and the Middle East with the help of local people, their lodges and their hospitality. Wildland Adventures donates its time and money to local organizations and causes.

Costa Rica

ECOFOCUS The environment and traditional culture of the southern Caribbean coast are threatened by overdevelopment of the tourism business by foreign agencies and by multinational corporations that are expanding their banana plantations.

RIO CAHABON JOURNEY
Raft in Guatemala with Maya Expeditions

Feel your heart pump and the adrenaline rush through your body as you plummet down Guatemala's best whitewater river, the Rio Cahabon. Class III and IV rapids alternate with calm stretches, allowing you to watch the rainforest pass by. On your way down the river, scan for iguanas and listen to monkeys and birds chatter. If you look closely at the branches supporting the deep jungle canopy, you may see the brilliant blues, yellows and reds of toucans and parrots, or perhaps you will catch a glimpse of the great blue heron diving for its dinner. Cave entrances covered with jungle growth line the shores; Mayas believed these to be entrances to the underworld. El Paradiso, an ancient natural thermal spring, offers therapeutic relaxation to you and fellow travelers. The complete five-day excursion visits the Quetzal Reserve and the ancient Mayan ruins of Quirigua, home to the largest monolith in Mesoamerica. Your adventure down the river will be as unique and spectacular as the rainforest itself.

ABOUT THE TRIP

LENGTH OF TRIP: 1, 2, 3 or 5 days
DEPARTURE DATES: Nov. to June
TOTAL COST: 5-day: $460; 3-day: $315
TERMS: Payment in US dollars only; 50% deposit; remainder due 30 days prior to departure; 30-50% cancellation fee
DEPARTURE: Arrive in Guatemala City; can provide a shuttle from airport
AGE/FITNESS REQUIREMENTS: 8-12 years minimum
SPECIAL SKILLS NEEDED: No special skills needed; rafting skills will be taught
TYPE OF WEATHER: Mild and warm with chance of rain; 70°-90°
OPTIMAL CLOTHING: Raingear, river shoes, mosquito protection

GEAR PROVIDED: All life jackets, paddles, rafting gear, tents, sleeping pads and food
GEAR REQUIRED: Flashlight
CONCESSIONS: Film and personal items available in Guatemala City
WHAT NOT TO BRING: Excess clothing or baggage
YOU WILL ENCOUNTER: Indigenous peoples in their villages; jungle river birds and wildlife
SPECIAL OPPORTUNITIES: Caving, hiking, swimming in hot springs, birdwatching
ECO-INTERPRETATION: Guides provide information on local flora, fauna, indigenous people and Mayan history.
RESTRICTIONS: No nude bathing
MEALS: Vegetarian meals are available upon request.

ABOUT THE COMPANY

MAYA EXPEDITIONS
15 Calle 1-91 Zona 10, Edificio Tauro #104, Guatemala City, Guatemala, (5022) 374-666
Maya Expeditions is a five-year-old for-profit company, owned by Tammy Ridenour and employing three Guatemalan guides in its staff of 7 to 12. The company puts most of its money back into the country and promotes low-impact ecotourism. Guide-training courses are offered to any interested citizen of Central America. The company also offers a student exchange program with scholarships for those who qualify.

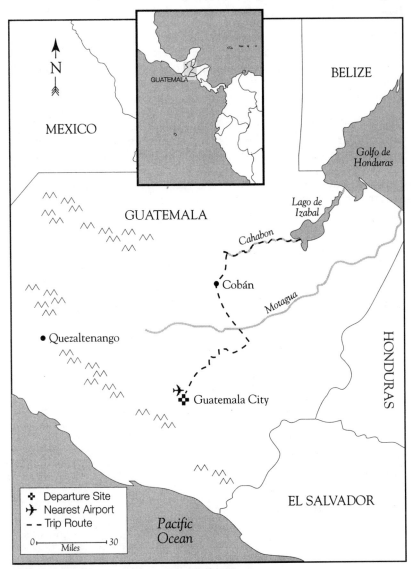

ECOFOCUS Deforestation ran rampant during the last decade as land was cleared to grow corn. Environmental groups have begun protesting the cutting of hardwoods. Their efforts have paid off, and deforestation has slowed in recent years, although the problem is far from solved. Also, construction of hydroelectric dams is being considered.

COPPER CANYON TREK
Backpack in Las Barrancas del Cobre with Adventure Associates

Copper Canyon, or Las Barrancas del Cobre, is 1,500 feet deeper than the Grand Canyon and encompasses an area larger than 20,000 square miles. It is known as the most spectacular canyon system in all of North America. The morning train to the small mining town of Creel, on the Chihuahua Pacific railway, will give you access to Basaseachi Falls, the highest waterfall in Mexico. You will visit the historic village of Batopilas, on the canyon floor, in the middle of Copper Canyon's tropical-fruit-producing region. The group also visits the lost cathedral of Satevo, believed to have been built in the 1600s, now the center of the community. Six days of backpacking along the canyon's rugged trails and rocky terrain gives hikers the opportunity to visit the Tarahumaran indigenous peoples, a private, seminomadic tribe. Days of hiking and nights of camping lead travelers to Urique, where transportation is provided to the train in Bauichivo, a one-way ride to Los Mochis, the endpoint of your journey.

ABOUT THE TRIP

LENGTH OF TRIP: 15 days
DEPARTURE DATE: December 1992
TOTAL COST: $1,285
TERMS: Checks accepted; $400 deposit with $100 nonrefundable
DEPARTURE: Nearest airport and departure from Los Mochis
AGE/FITNESS REQUIREMENTS: Good physical condition
SPECIAL SKILLS NEEDED: Experience in backpacking
TYPE OF WEATHER: Mild, dry days and cool evenings; low 55°-65°, high 75°-85°
OPTIMAL CLOTHING: Sturdy hiking boots and back-country clothing
GEAR PROVIDED: Tarps, cooking utensils, food and water pumps
GEAR REQUIRED: Tents, sleeping bags and pads

CONCESSIONS: Film can be purchased en route to the trek.
YOU WILL ENCOUNTER: Indigenous peoples (Tarahumara); numerous species of birds
SPECIAL OPPORTUNITIES: Photo opportunities of the rugged, spectacular and culturally rich canyon
ECO-INTERPRETATION: Travelers participate in discussions about canyon and desert ecosystems and visit with the Tarahumara Indians.
RESTRICTIONS: Backpackers must be aware that the native Tarahumara Indians are very private people.
REPRESENTATIVE MENU: *Breakfast*—cereal, fruit, eggs, bread; *Lunch*—cheese, crackers, fruit, soup; *Dinner*—pasta, tortillas, beans, rice, vegetables, pizza, bread, fruit; *Alcohol*—not provided or encouraged

ABOUT THE COMPANY

ADVENTURE ASSOCIATES
P.O. Box 16304, Seattle, WA 98116, (206) 932-8352
Adventure Associates and its local guides lead small group tours to remote canyons and areas in Africa, Central America and the Pacific Northwest. Trips focus on cross-cultural and wilderness experiences. The company practices low-impact camping, employs and supports local people and businesses and makes contributions to local community environmental or cultural organizations. In the last four years, its ecotours have promoted care and respect for the natural world while creating a profitable business for the owners, Cris Miller and Sandy Brown.

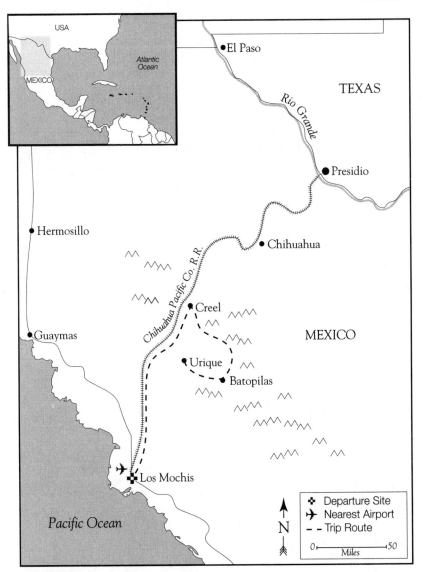

ECOFOCUS Copper Canyon provides a home for the Tarahumara indigenous peoples, who are currently being threatened by encroachment on their land by logging roads and edicts from the government forcing them to change their nomadic lifestyle to one of farming the canyon floor.

EL TRIUNFO CLOUD FOREST EXPEDITION
Birdwatch in Mexico with Victor Emanuel Nature Tours

The journey into the El Triunfo cloud forest is like a journey back in time. This pristine habitat provides a home for trees up to 100 feet high, enormous ferns and countless species of birds. Your small group will fly from Cancun to Tuxtla Gutiérrez where you will visit the Miguel Alvarez del Toro Zoological Park, filled only with animals native to Chiapas. From there you travel to Finca Prusia, about 4,000 feet above sea level. This marks the beginning of your hiking journey to El Triunfo, another 2,000 feet up. Along the way, you may see resplendent quetzals flying gracefully through the air while other rare bird species call out to one another from the canopy of the cloud forest. The sweet songs of the yellow grosbeak, the brown-backed solitaire and the black robin will lull you to sleep after your long hike through the forest. Next, you'll travel to Canada Honda, or Deep Canyon, home to the azure-rumped tanager. From there the group follows a trail through beautiful territory to the tropical lowlands, where opportunities abound to quench your birding thirst before returning to Tapachula and Mexico City, the end of your journey.

ABOUT THE TRIP

LENGTH OF TRIP: 11 days

DEPARTURE DATES: March of each yearpp

TOTAL COST: $1,992

TERMS: VISA, MC, AmEx; $300 deposit; cancellation fees vary depending upon the amount of advance notice

DEPARTURE: The trip originates at the airport in Cancun.

AGE/FITNESS REQUIREMENTS: None

SPECIAL SKILLS NEEDED: Ability to hike 4-6 hours per day at high elevations

TYPE OF WEATHER: 35°-90°, cooler at night, chance of rain

OPTIMAL CLOTHING: Sturdy hiking boots, raingear, hat and warm jacket

GEAR PROVIDED: Tents and tarps available to rent, cooking equipment

GEAR REQUIRED: Sleeping gear, eating utensils, day pack, water bottle and a flashlight

CONCESSIONS: Film and personal items may be purchased in towns near the trails.

WHAT NOT TO BRING: Nice clothes, unnecessary items

YOU WILL ENCOUNTER: Opportunities to view the quetzal, horned and highland guan, black-throated jay and blue-throated motmot and their breeding spots

SPECIAL OPPORTUNITIES: Photographing local people, beautiful scenery and wildlife

ECO-INTERPRETATION: Tour leaders will discuss ecosystems and help to identify birds, trees, mammals and reptiles.

RESTRICTIONS: None

REPRESENTATIVE MENU: *Breakfast*—granola; *Lunch*—tuna; *Dinner*—chicken, some meat, beans, rice, tomatoes; *Alcohol*—not included; bring or buy your own

ABOUT THE COMPANY

VICTOR EMANUEL NATURE TOURS
P.O. Box 33008, Austin, TX 78764, (512) 328-5221, (800) 328-VENT

Sixteen years ago Victor Emanuel was one of the pioneers of ecotourism.Victor Emanuel Nature Tours (VENT) promotes respect of the natural world through education and experience. Naturalist guides lead birdwatching and natural history tours throughout the world. VENT is a for-profit company that donates $100 per participant of this tour to the Institute of Natural History of Chiapas, a regional agency that protects cloud forests and natural areas. VENT also provides an added income for the local economy, creating an incentive for the locals to fight deforestation in their area.

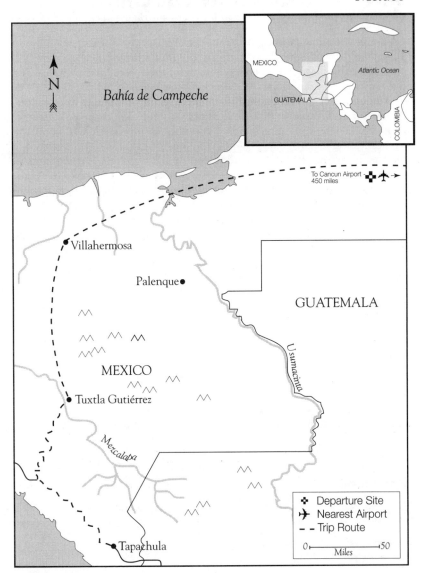

Mexico

ECOFOCUS Deforestation has slashed and burned ancient rain- and cloud forests throughout Mexico and Central America, stripping exquisite forests of their wildlife. This threat endangers El Triunfo and the few remaining areas like it.

SAN IGNACIO WHALE WATCHING
Whale watching in the San Ignacio Lagoon with Baja Expeditions

A long their migration path from the Arctic, the Pacific gray whales swim through three lagoons along the west coast of the Baja Peninsula. San Ignacio Lagoon is the most remote of the three. Its mangrove-lined estuaries and low dunes provide a surreal backdrop from which to view this enormous mammal. From your first moment at the Whalewatch Basecamp, American and Mexican guides, skiff drivers and cooks ease you through your whale-watching experience. You will spend three days watching whales from the skiffs. The other two days you can participate in activities like sea kayaking through crystal waters or hiking along fragile dunes in sculpted sand. You may prefer to observe the diverse organisms of the mangrove estuaries or comb the beach for sand-covered treasures. There are also many natural history lectures available to help answer your questions. On this tour, you will practice low-impact camping and benefit the Baja economy by supporting local businesses. This exposure to the magnificent Pacific gray whale and many other extraordinary species of wildlife in this area will increase your ecological awareness, while providing you with countless exciting adventures.

ABOUT THE TRIP

LENGTH OF TRIP: 5 days

DEPARTURE DATES: Jan.-Mar.

TOTAL COST: $1,275

TERMS: Accept all types of payment; $300 deposit; cancellation fees dependent upon length of time before trip

DEPARTURE: Van and airplane transportation to San Ignacio furnished by Baja Expeditions from San Diego, the nearest airport and point of departure.

AGE/FITNESS REQUIREMENTS: 8 years and older; general good health

SPECIAL SKILLS NEEDED: None

TYPE OF WEATHER: Mornings are foggy and cool; 70° in the afternoons; 50-60° in the evenings

OPTIMAL CLOTHING: Waterproof shoes, swimwear, light clothing for daytime and warm clothing for nights

GEAR PROVIDED: Tents, cots, sleeping bags, skiffs and food

GEAR REQUIRED: Water bottle, binoculars, flashlight, sunglasses, sunscreen

CONCESSIONS: Purchase film and personal items before the trip.

WHAT NOT TO BRING: Items that you mind getting wet

YOU WILL ENCOUNTER: Local inhabitants of San Ignacio, whales, dolphins, sea lions, cormorants and other birds

SPECIAL OPPORTUNITIES: Kayaking, hiking, birdwatching, beachcombing

ECO-INTERPRETATION: Your naturalist guide will provide information on the history of the Baja Peninsula as well as its bird life and whale population. There are also slide shows, lectures and guidebooks available.

RESTRICTIONS: Must follow regulations of Vizcaino Biological Reserve

REPRESENTATIVE MENU: *Breakfast*—cereal, pancakes, fruit; *Lunch*—ceviche, sandwiches; *Dinner*—chicken, seafood, carne asada; *Alcohol*—beer and wine supplied with dinner

ABOUT THE COMPANY
BAJA EXPEDITIONS
2625 Garnet Ave., San Diego, CA 92109, (619) 581-3311, (800) 843-6967

Baja Expeditions, a 17-year-old, for-profit company, staffs 12 full-time employees and 40 guides in its attempt to promote conservation and awareness through ecologically sound adventure travel to Baja, Mexico and Costa Rica. Kent Madin and Tim Means' profitable business promotes conservation of natural environment and habitat in addition to contracting with local businesses to support local economies.

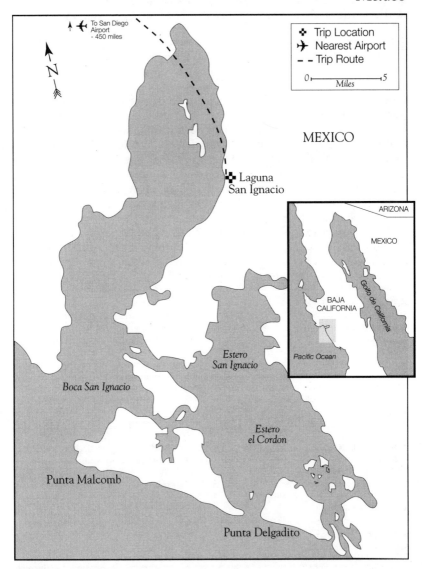

To San Diego Airport - 450 miles

N

❖ Trip Location
✈ Nearest Airport
– – Trip Route

0 |———————| 5
Miles

MEXICO

■ Laguna San Ignacio

ARIZONA

MEXICO

BAJA CALIFORNIA

Golfo de California

Pacific Ocean

Estero San Ignacio

Boca San Ignacio

Estero el Cordon

Punta Malcomb

Punta Delgadito

ECOFOCUS Baja's pristine west coast is threatened by excessive, insensitive tourism. Baja Expeditions has pioneered efforts to control access to the gray whales. Permits are distributed to only three companies during the whale-watching season. They discourage unregulated whale-watching, which can result in danger to humans and whales if the whales feel threatened.

USUMACINTA RIVER ADVENTURE
Raft the "River of the Sacred Monkey" with Slickrock Adventures

Rafting down the Usumacinta River, one of the largest rivers in Mexico, through the second-largest rainforest in the world, might be enough for some adventure travelers, but Slickrock Adventures takes you a few steps further. Travel through some of the wildest jungle in Mexico, to ancient ruins accessible only by rafting down the "River of the Sacred Monkey," also called the "Usu" and the "River of Ruins," which forms the border between Mexico and Guatemala. This downstream journey allows entrance to three Mayan cities: Yaxchilan, a restored city famous for its lintels, stone stelae and the Temple of the Bird Jaguar; Piedras Negras, an unrestored city known for its pyramids covered with jungle overgrowth and the sacrificial rock at the ruin gateway; and, finally, El Cayo, an unrestored city with extensive ruins to explore. You may want to take some time to wander about these ruins left by a mysterious race, or you can improve your kayaking under the watchful eyes of your guides, view the abundant wildlife, swim in waterfalls or make camp on one of the large, sandy beaches. No matter what you prefer, you will find a balance between serenity and excitement during this unique adventure.

ABOUT THE TRIP

LENGTH OF TRIP: 10 days
DEPARTURE DATES: Jan.-Mar.
TOTAL COST: $1,200
TERMS: Checks; $200 deposit reserves space, balance due 45 days before trip; full refund minus $50 fee for cancellations at least 60 days in advance
DEPARTURE: A charter van provides transportation from the airport in Villahermosa, Mexico, to the river.
AGE/FITNESS REQUIREMENTS: 10 and over, good physical condition
SPECIAL SKILLS NEEDED: Swimming ability
TYPE OF WEATHER: Hot and humid, rain possible
OPTIMAL CLOTHING: Tennis shoes, polypropylene, long-sleeved shirts for warmth
GEAR PROVIDED: All river and kitchen gear
GEAR REQUIRED: None

CONCESSIONS: Neither film nor personal items will be available on the river.
WHAT NOT TO BRING: Excessive clothing
YOU WILL ENCOUNTER: Some local people, howler monkeys, parrots, toucans, jaguars, deer and other indigenous species
SPECIAL OPPORTUNITIES: Exploring ancient Maya ruins, hiking in the jungle and swimming in waterfalls
ECO-INTERPRETATION: Guides provide information about jungle ecology and Maya archaeology.
RESTRICTIONS: Must respect the local people
REPRESENTATIVE MENU: *Breakfast*—eggs and toast; *Lunch*—turkey and cheese sandwiches; *Dinner*—chutney chicken, parmesan rice, fresh vegetables; *Alcohol*—bring your own or purchase before trip in Palenque

ABOUT THE COMPANY

SLICKROCK ADVENTURES
P.O. Box 1400, Moab, UT 84532, (801) 259-6996
Cully Erdman has owned and operated Slickrock Adventures for 15 years. Professional guides specializing in outdoor skills and familiar with their specific region lead adventure-seeking travelers through parts of Utah, Mexico and Belize. These low-impact, non-motorized journeys instruct people in outdoor skills necessary for river navigation, mountain biking and ecologically sound waste disposal. Slickrock is a member of the Southern Utah Wilderness Alliance, a local rainforest protection organization in Belize, and cleans up cays on Belize's barrier reef.

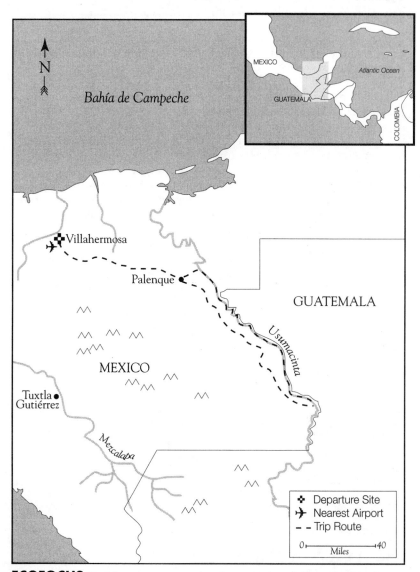

ECOFOCUS The Mexican government has plans to construct a dam at San José Canyon on the Usumacinta. This would result in a lake that would back up for over 100 miles and drown the ruins of huge stone cities at Yaxchilan and Piedras Negras. The government appears likely to halt construction plans due to opposition. Deforestation is an enormous problem on the Mexican side of the river, while the rainforest on the Guatemalan side remains virtually untouched.

TRANS-DARIEN EXPEDITION
Tour tropical forests of Panama with Eco-Tours de Panama

This trip, led by Eco-Tours de Panama S.A., is a trek through the rainforest of the Darien Gap, one of the least-explored regions on Earth. Following close to the route used by Balboa when he "discovered" the Pacific Ocean in 1503, this 12-day expedition crosses from the Atlantic coast near the Colombian border to the town of Yavisa, not far from the Pacific Ocean. Each expedition has a maximum of 15 participants, plus an expedition leader, assistant leader, native guide and porters to carry the food and heavy equipment. Everything is included except personal clothing, cameras and binoculars, which can be carried in a small knapsack. Camp is made each day by the porters prior to the arrival of the main party. Each expedition includes three "rest" days when no travel is planned, allowing participants the leisure to thoroughly enjoy the highlights of this area. During this time, you can observe the magnificent jungle flora and fauna of this extraordinary rainforest. You will also be exposed to the fascinating customs of indigenous peoples through visits to the settlements of the Kuna and Chocoe/Embara Indians. The cost of the tour includes a seminar in jungle survival and a donation toward the maintenance of Panama's Darien National Park.

ABOUT THE TRIP

LENGTH OF TRIP: 12 days

DEPARTURE DATES: Jan.-June each year

TOTAL COST: Varies

TERMS: Accept US dollars in cash for the 50% advance deposit; cancellation policy is 10% after 30 days

DEPARTURE: From Panama City's airport

AGE/FITNESS REQUIREMENTS: Bring a doctor's certificate saying you are in good condition to walk.

SPECIAL SKILLS NEEDED: Fit and able to walk distances

TYPE OF WEATHER: Hot and humid throughout the year, a bit drier from Jan. to Mar.

OPTIMAL CLOTHING: Long pants and shirts with waterproof shoes for the jungle

GEAR PROVIDED: Everything except clothing and personal items

GEAR REQUIRED: Day pack

CONCESSIONS: Film and personal items can be purchased in Panama City prior to trip.

WHAT NOT TO BRING: Sleeping bags, hammocks or cumbersome gear

YOU WILL ENCOUNTER: Kuna Indian and Chocoe/Embera settlements, as well as various wildlife on the Trans-Darien

ECO-INTERPRETATION: Printed information is available upon arrival, guides are trained in the natural history of Panama and a survival course is included.

RESTRICTIONS: Respect for indigenous peoples and wildlife

REPRESENTATIVE MENU: *Breakfast*—cereals, fruit, juices; *Lunch*—sandwich while on treks, fruit; *Dinner*—pastas, potatoes, dried meat, rice and beans, lentils; *Alcohol*—none

ABOUT THE COMPANY

ECO-TOURS DE PANAMA S.A.
Apartado 465, Panama City 9A, Panama, (507) 363-575
This for-profit Panamanian business uses native guides and locally owned hotels. In addition to the "high" adventure Trans-Darien Expeditions, it also offers "soft" nature tours to Panama's rainforests bordering the Panama Canal and cloud forests (above 5,000 feet). The company promotes conservation of Panama's natural resources by providing each of its participants with a one-year membership in Ancon, one of Panama's nonprofit rainforest-conservation organizations. Ancon also receives support from the Nature Conservancy and the World Wildlife Fund.

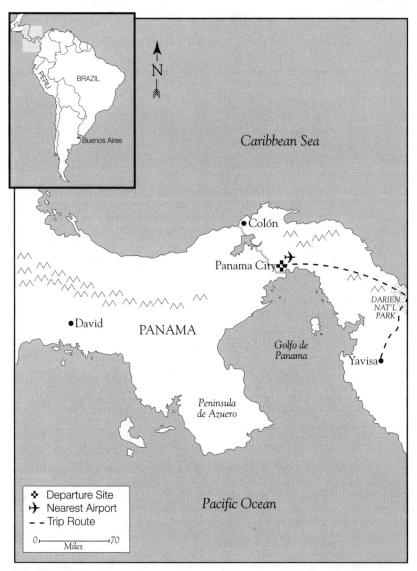

Caribbean Sea

PERU
BRAZIL
Buenos Aires

N

Colón

Panama City

DARIEN
NAT'L
PARK

David

PANAMA

Golfo de
Panama

Yavisa

Peninsula
de Azuero

❖ Departure Site
✈ Nearest Airport
- - Trip Route

0 ⊢————————⊣ 70
Miles

Pacific Ocean

ECOFOCUS Deforestation caused by expansion of grazing lands is a grave problem in Panama, endangering many species of flora and fauna. This country has one of the richest ecosystems in the world, with 1,200 species of birds, 1,500 species of trees and over 7,000 kinds of plants.

IGARAPÉ AND TERRE FIRME FOREST
Explore the rainforests of northern Brazil with Ecotour Expeditions

You are silently drifting down a narrow stream lined with flowering plants and liana vines. Time seems to stand still, as you float along listening to a chorus of birdcalls. The boat glides past large fruits dropped by surrounding trees. Suddenly, you hear a loud slap on the water. Before you can see where the sound came from, a fruit is jerked below by one of the many fruit-eating fish of the Negro River. With every passing minute, you behold more exotic species of plants and animals than you ever thought possible. You are on the Igarapé and Terre Firme Forest Expedition in Brazil. Ten days are spent exploring the Igarapés, tributaries of the Negro River, and hiking through the Terre Firme forest. This forest is filled with fascinating wildlife, including the screaming pia, the guaribas or howler monkeys, saúba or leaf-cutter ants, and tucandera ants. Throughout the trek, time is taken to identify plants integral to the lives of the native forest dwellers. Some vines contain a cool drink, others are strong enough to use as a rope. Clearly, the information is limitless, and so are the adventures on this journey through the wilds of Brazil. At the close of the trip, you will have two full days to tour the city of Manaus on your own time.

ABOUT THE TRIP
LENGTH OF TRIP: 15 days
DEPARTURE DATES: June 28, Aug. 16
TOTAL COST: $1,650
TERMS: Cash; $300 deposit; no penalty for cancellation up to 30 days before trip
DEPARTURE: Arrive in Manaus airport; further transport provided
AGE/FITNESS REQUIREMENTS: All ages; reasonably fit
SPECIAL SKILLS NEEDED: Curiosity and love of nature
TYPE OF WEATHER: Dry tropical equatorial climate; 83° days, 65° nights
OPTIMAL CLOTHING: Loose-fitting cotton clothing
GEAR PROVIDED: All forest camping equipment
GEAR REQUIRED: Sleeping bag, flashlight

CONCESSIONS: Buy film and personal items before the trip.
WHAT NOT TO BRING: Weapons
YOU WILL ENCOUNTER: Endless possibilities to view wildlife
SPECIAL OPPORTUNITIES: Catch-and-release fishing
ECO-INTERPRETATION: This trip is guided by highly acclaimed, accomplished biologists.
RESTRICTIONS: Follow the policy to leave indigenous peoples alone or to treat them with respect.
REPRESENTATIVE MENU: *Breakfast*—eggs, bread, coffee, juice; *Lunch*—sandwiches; *Dinner*—fish, beef or chicken, rice, salads, beans; *Alcohol*—wine available at additional cost

ABOUT THE COMPANY
ECOTOUR EXPEDITIONS
P.O. Box 1066, Cambridge, MA 02238, (800) 688-1822
The mission of Ecotour Expeditions is to explore the fabulously complex ecological relationships between plants and animals in tropical rainforests. It is a for-profit organization in business three years, with five staff people in the US and 15 in the field. All its trips are small-group, exploratory expeditions, often sighting new species of birds and plants. Local guides are hired and all waste is packed out.

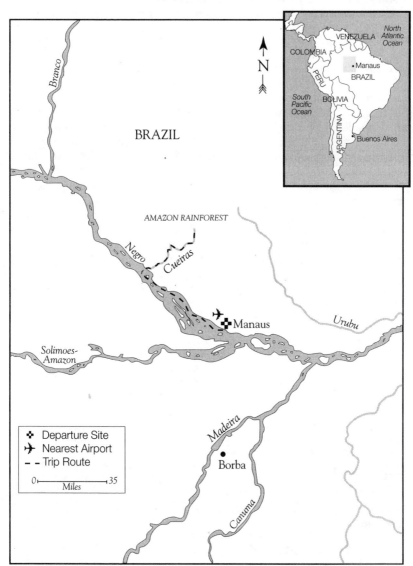

ECOFOCUS There are numerous endangered species in the rainforest, including the red uakari and the bearded saki primates, the harpy eagle and the peixe-boi, the Amazon manatee. Both the bearded saki and the harpy eagle are hunted for their meat. Deforestation is also a grave concern, as is the resulting danger to and displacement of indigenous peoples.

WHITE WATERS AND BLACK
Boat the Rio Negro and Rio Manacapuru with Ecotour Expeditions

This trip provides an in-depth exploration of the colorful Solimoes, Negro and Amazon rivers of Brazil. Aboard a large expedition vessel, the adventure begins with a visit to the igapo, a black-water system on the Negro. This "black water" is really tea-colored and high in acidity and decomposing leafy material. The trip ends in varzea, the whitewater-flooded forest of the Solimoes and Amazon, whose waters are grayish and muddy-colored by silt from the Andes. You will wake every morning to the sound of tropical birdcalls as they coordinate their feeding foray across the canopy. Throughout the day, you will board smaller boats with outboards to venture into the small streams or igarapés that wind deep into the forest. Hop out of the boat to explore whenever something sparks your interest. Every evening, the group gathers to talk about the day's discoveries and the complex ecological systems of the varzea and igapo. Nighttime is a good time to spy caimans lying quietly along the riverbanks fishing, their eyes glowing in the spotlights. Nocturnal birds like the nightjar are also abundant. Upon returning to Manaus, you will have a couple of days to enjoy this fascinating city by touring with the group or on your own time. Highlights include the markets, featuring various products extracted from the forest, major natural history museums and the Teatro do Amazonas, the famed opera house of the Amazon.

ABOUT THE TRIP

LENGTH OF TRIP: 15 days
DEPARTURE DATES: July 26, Sept. 20
TOTAL COST: $1,950
TERMS: Cash; $300 deposit; cancellation up to 30 days prior to trip
DEPARTURE: Arrive at Manaus airport; further transport provided
AGE/FITNESS REQUIREMENTS: None
SPECIAL SKILLS NEEDED: Curiosity and a sense of humor
TYPE OF WEATHER: Tropical rainstorms alternating with brilliant equatorial sun; 85° days, 65° nights
OPTIMAL CLOTHING: Loose-fitting cotton clothing
GEAR PROVIDED: All forest equipment

GEAR REQUIRED: Basic personal items
CONCESSIONS: Buy before the trip.
WHAT NOT TO BRING: Weapons
YOU WILL ENCOUNTER: Small pioneer villages; limitless wildlife
SPECIAL OPPORTUNITIES: Walk through ancient villages; visit contemporary villages of river people.
ECO-INTERPRETATION: The guides for this trip are accomplished biologists.
RESTRICTIONS: Treat natives with respect; follow the policy to leave indigenous peoples alone.
REPRESENTATIVE MENU: *Breakfast*—eggs, toast, yogurt, juice; *Lunch*—ham, chicken, salad, rice; *Dinner*—fish, beef, chicken, rice, potatoes, salads, fruit juice, fruits; *Alcohol*—available at extra cost

ABOUT THE COMPANY

ECOTOUR EXPEDITIONS
P.O. Box 1066, Cambridge, MA 02238, (800) 688-1822
The mission of Ecotour Expeditions is to explore the complex ecological relationships between plants and animals in tropical rainforests. It is a for-profit organization in business three years, with five staff people in the US and 15 in the field. All its trips are exploratory expeditions, often sighting new species of birds and plants. Local guides are hired, all waste is packed out and money is donated to scientific research.

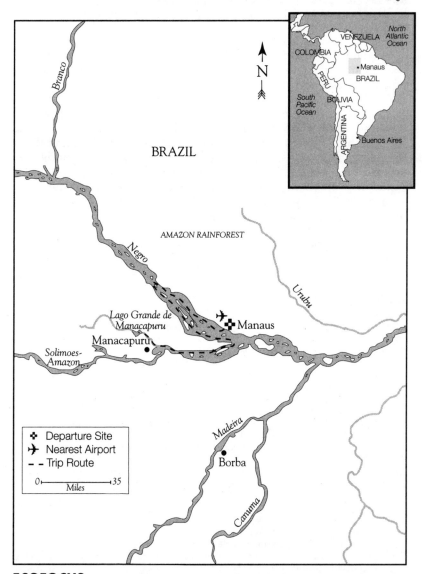

Brazil

ECOFOCUS Areas visited are too remote to be threatened by settlement or exploitation. However, it is crucial to keep these areas safe from those problems in the future. Deforestation and species extinction both of animals and endemic plants is an impending tragedy.

EASTER ISLAND EXPEDITION
Explore Easter Island with Nature Expeditions International

Referred to as Te Pito O Te Huenua, the navel of the world, Easter Island is the most isolated inhabited island on Earth. After an afternoon flight from Santiago, Chile, Sergio Rapu, archaeologist and leading expert on Easter Island history, will give an orientation lecture on local culture and archaeology. Easter Island is famous for its stone statues with elongated heads and ears, representing ancestral men called Moais. Among the many interesting sites to visit are Anakena Bay, the first settlement on the island, Ahu Nau Nau, a restored site overlooking the bay, and Rano Raraku, a mysterious and fascinating quarry filled with statues. You will also visit the ceremonial village of Orongo, located at the edge of a volcanic crater, and ruins along the south coast, including Ponkura, Vauhu and Akanhanga. Day trips will be taken to the archaeological zone, Ahu Tahai, the Easter Island Museum and Ahu Akivi, a restored group of statues. You will have ample opportunity to join the islanders in daily activities such as dancing and singing, a Polynesian feast, fishing for tuna, horseback riding, swimming, schooling and religious services, before you depart on an early morning flight back to Chile.

ABOUT THE TRIP

LENGTH OF TRIP: 15 days

DEPARTURE DATES: Feb., June, Oct., Dec.

TOTAL COST: $2,190

TERMS: Check; $300 deposit; cancellation fee of $50 with 60 days' advance notice

DEPARTURE: Arrive in Santiago, Chile, and transfer with group to Hotel Conquistador

AGE/FITNESS REQUIREMENTS: Good health; ability to walk 1-4 miles daily

SPECIAL SKILLS NEEDED: None

TYPE OF WEATHER: Breezy, cool weather with a chance of rain; 70° days, 40° nights in Feb., Oct. and Dec.; 60° days, 40° nights in June

OPTIMAL CLOTHING: Comfortable walking clothes and shoes

GEAR PROVIDED: None necessary

GEAR REQUIRED: Windbreaker

CONCESSIONS: No film or personal items available except in Santiago; souvenirs available from local artists

YOU WILL ENCOUNTER: Indigenous people; Polynesian ceremonies

SPECIAL OPPORTUNITIES: Horseback riding

ECO-INTERPRETATION: Discussion of the impact of tourism

RESTRICTIONS: Guides will discuss showing respect for cultural artifacts upon arrival.

REPRESENTATIVE MENU: *Breakfast*—coffee, tea, eggs, bread, fruit; *Lunch*—seafood, bread, fruit; *Dinner*—seafood, bread, vegetables; *Alcohol*—available but bringing your own is recommended

ABOUT THE COMPANY

NATURE EXPEDITIONS INTERNATIONAL
P.O. Box 11496 , Eugene, OR 97440, (503) 484-6529 (800) 869-0639

For 19 years with a staff of five, Nature Expeditions International has been leading natural history and cultural expeditions around the world. Started as a nonprofit organization, Nature Expeditions International became a for-profit company after ten years. Presently, they donate to Oregon conservation groups and fund three environmental conservation scholarships. They use local guides and lodging.

MEXICO

Caribbean Sea

N

VENEZUELA

COLOMBIA

Galápagos

BRAZIL

PERU

BOLIVIA

CHILE

Rio de Janeiro

Easter Island

Santiago

ARGENTINA

South Pacific Ocean

South Atlantic Ocean

Falkland Islands

✤ Departure Site
✈ Nearest Airport
- - Trip Route

0 ⊢————————⊣1000
      Miles

ECOFOCUS Current threats to cultural sites come from poorly supervised tour groups scoring or scraping existing designs with stones to make them more visible for photographs. The local ecology is presently preserved, but conservation measures are essential to protect the natural resources of the area from the potential effects of overpopulation, like deforestation which nearly destroyed the island 400 years ago.

LAND OF THE PEHUENCHE
Explore the mountains of Chile with Andes Trekking

This Andean experience is divided into three sections, each one week long. The first is devoted to environmental awareness. Participants will camp on private property in the mountainous outskirts of Santiago, taking day hikes and being briefed by environmental leaders. This area is a virtual playground of spectacular waterfalls, glimmering glaciers, hot rock climbing spots and rugged terrain hot springs—all just 90 minutes from the coast. There are more than 300 canyons within a 100-mile radius just waiting to be explored. The next week, entitled "living earth week," is the core of this expedition and is dedicated to developing communion with the environment. During this time, there will be group and individual activities ranging from breathing techniques and terrain analysis to walking with awareness and rope handling. The final week focuses on sharing ancient life ways with the native Pehuenches, Araucanos or Diaguitas, depending on the season. Andes Experience staff believes that this exposure to the local native and urban cultures, as well as to the magnificent wilderness of the Andes, will provide participants with valuable insight. This knowledge is crucial to the development of a sustainable society, which is so necessary in the world today.

ABOUT THE TRIP

LENGTH OF TRIP: 14, 21 and 25 days
DEPARTURE DATES: 1st and 15th of every month
TOTAL COST: $1,600-$2,800
TERMS: Checks; 25% deposit; cancellation up to 35 days prior
DEPARTURE: Arrive in Santiago.
AGE/FITNESS REQUIREMENTS: All ages; good fitness
SPECIAL SKILLS NEEDED: None
TYPE OF WEATHER: Desert: 35°-95° and dry. Rainforest: 45°-80° and damp
OPTIMAL CLOTHING: Depends on the area and season (you will be informed in advance)
GEAR PROVIDED: Tents, ropes, radio equipment
GEAR REQUIRED: Mountain boots, internal-frame pack, canteen, sleeping bag, windbreaker anorak, high-quality sunglasses, folding knife

CONCESSIONS: Film, souvenirs and personal items can be bought in Santiago.
YOU WILL ENCOUNTER: Funny guides, environmentalists, natives, urban Bohemians, goats
SPECIAL OPPORTUNITIES: Mountain trekking, rock climbing, horseback riding
ECO-INTERPRETATION: A whole week is designed for communication with "hand-on environmental organizations."
RESTRICTIONS: No radios, smoking, or drugs
MEALS: In urban areas: fine local restaurants; Chilean gourmet vegetarian food available

ABOUT THE COMPANY

ANDES TREKKING
12021 Wilshire Blvd., Los Angeles, CA 90025, (310) 575-3990
Andes Experience is a company run by Chilean mountain guides and environmental activists. Its main focus is to offer a full experience of the Andean natural environment and culture, as well as an awareness of the relationships that affect the future of both. The company's motto is: "Fun and awareness are not mutually exclusive—they can be complementary." Director Adolfo Aguirre has run mountain, archaeological and rescue expeditions for US universities and private parties for eight years.

ECOFOCUS The primary environmental problem in this area is the displacement of indigenous peoples along the Bío-Bío River due to proposed dams. The first foundations of the dams have already been built, and more are being proposed. The Pehuenche and many other native populations would be forced to give up their land to make way for the flooding, and it is likely their cultures would not survive.

PATAGONIA EXPLORER
Expedition through Chilean Patagonia with Overseas Adventure Travel

"Variety" is the word to remember when explaining Overseas Adventure Travel's Patagonia Explorer expedition. Not only are your sights and experiences endlessly diverse, but so are your means of transport. Camp for three nights in the magnificent Paine Towers National Park amidst meadows bursting with wildflowers, where you can see condors, guanacos, Andean foxes, rheas and much more. Then hop aboard rubber rafts and descend the Grey and Serrano rivers to the Pacific. Here you can indulge in sun, rafting in remote wilderness beneath the South Patagonia ice cap, wildlife watching and visiting remote ranches. Finally, spend four days cruising the lovely forest-clad fjords of the Chilean archipelago in a comfortable motor-sailer. See such marine wildlife as petrels, albatross, sea lions, penguins and perhaps orcas. Meals feature delicious Chilean fruits, vegetables and fresh seafood. Your leader will be Jack Miller, who has been exploring the ice cap and remote fjords of the southern Andes since 1964. After a lifetime of mountaineering, rafting and trekking to lost Inca cities, Patagonia remains his favorite wilderness haven. The Patagonia Explorer provides an adventure-travel trip filled with excitement and diversity while being kind to the environment.

ABOUT THE TRIP

LENGTH OF TRIP: 16 days

DEPARTURE DATES: Jan. 18, Feb. 8, 29, Nov. 21, Dec. 19

TOTAL COST: $2,895

TERMS: Checks; $300 non refundable deposit payable by credit card

DEPARTURE: Arrive in Santiago airport; fly that night to Punta Arenas.

AGE/FITNESS REQUIREMENTS: All ages; good health proven by medical form

SPECIAL SKILLS NEEDED: None

TYPE OF WEATHER: Sunny or cloudy and cool, rainy and windy; 30°-70° but wind can make it much colder

OPTIMAL CLOTHING: Layers, hiking boots, polypropylene bunting jacket, raingear

GEAR PROVIDED: All group camping equipment

GEAR REQUIRED: Day pack, personal clothing, sleeping bag and pad

CONCESSIONS: Bring film from home; souvenirs are available.

YOU WILL ENCOUNTER: Local ranchers; opportunity to view guanacos, condors, foxes, black-necked geese, penguins and seabirds

SPECIAL OPPORTUNITIES: Endless landscapes and glaciers for photographing

ECO-INTERPRETATION: Guides provide ongoing information on indigenous wildlife, geology, etc.

RESTRICTIONS: None

REPRESENTATIVE MENU: *Breakfast*—eggs, hot chocolate, coffee, toast, jam, tea, fresh fruit; *Lunch*—sandwiches with cold cuts, fresh fruit; *Dinner*—chili, wine, fresh soup, chicken dish with veggies, flan for dessert; *Alcohol*—available for purchase

ABOUT THE COMPANY

OVERSEAS ADVENTURE TRAVEL
349 Broadway, Cambridge, MA 02139, (617) 876-0533, (800) 221-0814
Since 1978, Overseas Adventure Travel (OAT) has been providing trips that are sensitive to the environment, led by guides who respect the land and its people and pass along these important values to their travelers. OAT employs mostly local guides, camps where appropriate, and practices waste-disposal and fuel-use restrictions. OAT is a for-profit organization, owned by Judi Wineland, offering travel all around the world.

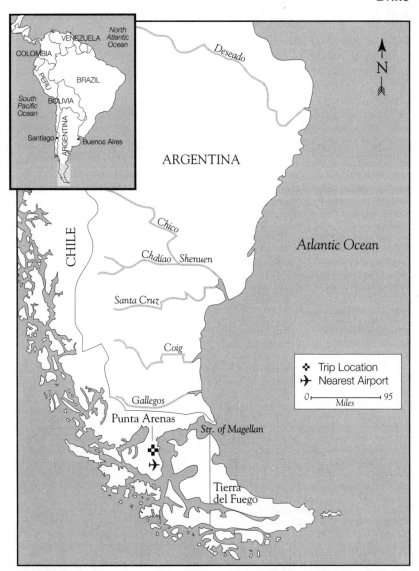

ECOFOCUS The area suffers from dilemmas about overfishing of king crab and shellfish, and overharvesting of marine resources. Deforestation is a problem in southern Chile. Puma are disappearing because of conflicts with farmers over the killing of sheep. Guanacos, once endangered, are now protected and flourish in Paine Towers National Park.

RIO BÍO-BÍO WHITEWATER
Raft the Bío-Bío in Chile with Nantahala Outdoor Center

"Bío" is a Mapuche Indian word meaning "water". When repeated, it means lots of water. For eight days, you will ride huge waves that crash through lava-rock canyons, past tumultuous tributaries, limestone spires, hot springs and smoking volcanoes. Ask anyone who has been to this extraordinary river—rafting the Bío-Bío is the ultimate whitewater experience. The river is also part of a magnificent wilderness area that could disappear completely in the near future if up to six proposed dams are accepted and funded. By participating in this raft trip with Nantahala Outdoor Center, you can be assured that a portion of your fee will serve as a financial contribution to the preservation of this extraordinary environment. Highlights of this trip include the Volcano Callaqui, which stands over 10,000 feet in elevation and continues to smoke, many species of birds–buff-necked ibises, ashy-headed canquen geese and kingfishers–and "The Canyon of 100 Waterfalls." Before the raft trip, you will have time to explore Santiago and the enchanting foothills of the Andes, including the town of Loquimay, where horses are hitched outside taverns and oxen pull carts through the streets. Before going home, you will visit the open-air market of Chillán, which offers an assortment of native handicrafts.

ABOUT THE TRIP

LENGTH OF TRIP: 13 days
DEPARTURE DATES: Jan. 9, 23, Feb. 6
TOTAL COST: $2,100
TERMS: Check or credit card; $250 deposit; 60 days cancellation policy
DEPARTURE: Travelers will be met at the airport in Santiago by NOC guides.
AGE/FITNESS REQUIREMENTS: Must be at least 14 years old; in good health
SPECIAL SKILLS NEEDED: Experience in rafting; expert skill in kayaking and canoeing
TYPE OF WEATHER: Temperate; 75°-80° in the day, 50°-60° at night; occasional rain
OPTIMAL CLOTHING: Hiking shoes, long pants, visor or hat, gloves, raingear, jacket or sweater, polypropylene, paddling jacket and/or wet suit
GEAR PROVIDED: Water-resistant river bags
GEAR REQUIRED: Sleeping bag, ground cloth, sleeping pad, pack.

CONCESSIONS: Film can be purchased, but is more expensive
WHAT NOT TO BRING: Excess gear
YOU WILL ENCOUNTER: Indigenous peoples (Mapuche), wildlife
SPECIAL OPPORTUNITIES: Hiking
ECO-INTERPRETATION: All NOC instructors are professional whitewater guides.
RESTRICTIONS: Treat local people with respect.
REPRESENTATIVE MENU: *Breakfast*—fresh fruit, French toast, pancakes, omelettes, yogurt, cereal, coffee; *Lunch*—bread, deli meats, salads, cheese, vegetables, fruit, chips and crackers; *Dinner*—salad, vegetables, bread, leg of lamb or salmon, traditional cabrito asado (roast beef), desserts; *Alcohol*—Chilean wines are provided with dinner

ABOUT THE COMPANY
NANTAHALA OUTDOOR CENTER
41 Hwy. 19 West, Bryson City, NC 28713, (704) 488-2175
Recipient of the Corporate Achievement in River Conservation Award presented by American Rivers, Nantahala Outdoor Center is an extremely active environmental leader. It is a for-profit company managed by Bob Powell and employee-owned. It has been in business 20 years and specializes in whitewater kayaking, canoeing and rafting trips with a staff of 400 during peak season. Profits are donated to a number of conservation organizations, among which are NOC Environmental and Humanitarian Fund, The Tides Foundation, Fund for the Preservation of the Bío-Bío, and the Outdoor Industry Conservation Alliance.

ECOFOCUS The Chilean government plans to build as many as six dams along the Bío-Bío, one of Chile's largest rivers. This would provide the country with much-needed jobs and an efficient source of electricity. It would also flood 52,000 acres along the river, endangering a fragile rainforest environment, and many species of wildlife and bringing an end to traditional indigenous Pehuenche life. One of the dams, the Pangue, has already been approved. Five more remain under consideration.

DISCOVERING CUYABENO WILDLIFE

Jungle adventure in Ecuador with Southwind Adventures

The ecosystems of the Amazon Basin are some of the richest and most complex in the world. In Ecuador, up to 100 species of trees have been recorded within one acre of forest, as compared to 40 species per acre in Central America and 20 species per acre in North America. Within the dense green walls of this jungle live 1,000 species of birds, 600 species of fish, 4,500 species of butterflies and 250 species of reptiles and amphibians, plus 10 species of monkey, tapir, capybara and jaguar, among countless other mammals. Here is your opportunity to experience the ultimate in bio-diversity as you travel by foot, motorboat and canoe. Your trip will explore Ecuador's Oriente, a territory east of the Andes containing everything from lush cloud forests to low-elevation jungles. You will board motorized canoes in Lago Agrio and ride for four days along the Aguarico and Cuyabena Rivers, meet local Cofan, Secoya and Siona Indians and observe the end-less variety of Amazon creatures. Spend three nights at the Cuyabeno Lodge in the heart of the Cuyabeno Wildlife Reserve learning about wildlife biology and ethnobotany with naturalist guides in the nearby virgin jungle during the day. Discovering Cuyabeno Wildlife Reserve is an extraordinarily colorful rainforest experience.

ABOUT THE TRIP

LENGTH OF TRIP: 12 days

DEPARTURE DATES: Feb.-Nov.

TOTAL COST: $1,695

TERMS: Checks, VISA, MC; $400 deposit; cancellation up to 45 days prior

DEPARTURE: Arrive at airport in Quito; private vehicle and charter plane provided for further transport

AGE/FITNESS REQUIREMENTS: Must be at least 8 years old and in good health

SPECIAL SKILLS NEEDED: None

TYPE OF WEATHER: Warm, humid with high possibility of short, heavy showers

OPTIMAL CLOTHING: Lightweight trousers, long-sleeved shirts, rubber boots

GEAR PROVIDED: All camping equipment, canoes and other vehicles

GEAR REQUIRED: Clothing and sleeping bag

CONCESSIONS: Film and personal items items can be purchased in Quito.

WHAT NOT TO BRING: Weapons, military-type clothing, loud radios

YOU WILL ENCOUNTER: Indigenous peoples daily, including the Cofan, Secoya and Siona Indians; camps are made in or near villages, and guides are natives; superb wildlife viewing

SPECIAL OPPORTUNITIES: Fishing

ECO-INTERPRETATION: Groups are accompanied by native Siona Indian naturalists and biologists who provide constant learning opportunities.

RESTRICTIONS: Treat native peoples with respect.

REPRESENTATIVE MENU: *Breakfast*—coffee, chocolate, jams, bread, eggs, tea, etc.; *Lunch*—salad, empanadas, sandwiches, juice, etc.; *Dinner*—fresh fish, chicken, rice, native fruits and vegetables, soups, etc.; *Alcohol*—available at the lodge and in the city only

ABOUT THE COMPANY

SOUTHWIND ADVENTURES
P.O. Box 621057, Littleton, CO 80162-1057, (303) 972-0701

A for-profit adventure and cultural-exploration company owned by Susie Shride, Southwind Adventures has been in business for three years in the US and 12 years in Peru. It gives money to Peru National Parks, ECO and APECO and is contributing to the cleanup of the Inca Trail. It also employs local guides who are specialists on traveling with minimum impact on the surroundings.

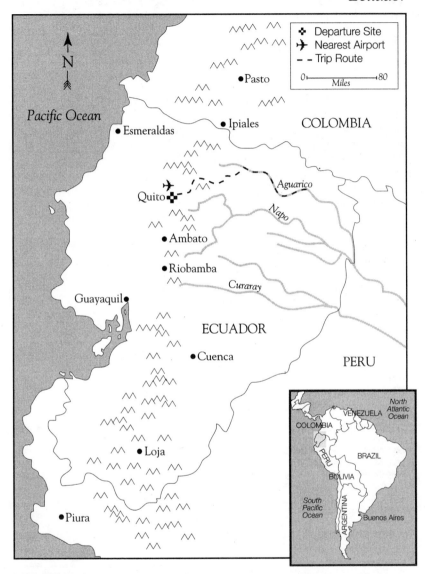

ECOFOCUS The rainforests in this area are extremely threatened by the conflicting views and needs of indigenous peoples, coupled with Ecuador's suffering economy. The Amazon Basin's forests are strained from hunting and slash-and-burn agriculture practiced by native peoples. This, however, does not occur in the Cuyabeno Wildlife Reserve. The Napa River area suffers from oil spills, pipeline leakage and people from the oil industry settling in the area.

THE INCA TRAIL TO MACHU PICCHU
Trek the Inca Trail in Peru with Terra Adventures, Inc.

The most dramatic way to arrive at the "Lost City of the Incas" is the way the Incas themselves did: by walking the magnificent paved trail of stones, stairways and tunnels winding across lofty ridges, seemingly as close to the sky as to the depths of the Urubamba Gorge far below. Interrupted by a series of imposing Inca ruins that dominate the landscape in all directions, the 27-mile royal road ascends to great heights, revealing sweeping views of towering snow-covered peaks, then descends into the cloud forest, where it more resembles a tunnel than a trail, so dense is the tropical vegetation. Beyond the final granite stairway to Inti Punku stands revealed in all its glory the masterwork of the Inca empire: Machu Picchu. It is the perfect union of the work of humanity and the natural environment, a city perched atop a narrow crest high above the Urubamba River, surrounded by the intense green of the rugged mountains of the Amazonian cloud forest. It is impossible to conceive of a more spectacular setting for the worship of the gods. With Terra Adventures, Inc., you can enjoy this overwhelming experience without a worry. Porters carry your gear, and delicious meals and spacious tents are provided. After the trek, you will return to Cuzco, the former Incan capital, by train.

ABOUT THE TRIP

LENGTH OF TRIP: 8 days

DEPARTURE DATES: The 2nd and 4th Fridays, Apr.-Oct.

TOTAL COST: $675

TERMS: Checks; $300 deposit; cancellation fee of $50, further charges for cancellations less than 30 days before trip; no refund for cancellations 7 days before trip

DEPARTURE: Arrive at Lima airport; drive to hotel in company cars.

AGE/FITNESS REQUIREMENTS: None

SPECIAL SKILLS NEEDED: None

TYPE OF WEATHER: Cold in highlands (50°-60° days, 0°-35° nights), warm in cloud forest (75°-85° days, 50°-70° nights)

OPTIMAL CLOTHING: Both warm and loose-fitting clothing, hiking footwear

GEAR PROVIDED: Tents, sleeping pads

GEAR REQUIRED: Rain poncho, sleeping bag, water bottle, flashlight

CONCESSIONS: Bring your own film; souvenirs and personal items are available in local markets.

WHAT NOT TO BRING: Formal clothing, hard luggage, candy for indigenous children

YOU WILL ENCOUNTER: Exposure to many indigenous peoples, magnificent birdwatching, including condors

SPECIAL OPPORTUNITIES: Birdwatching, photography

ECO-INTERPRETATION: Explanation about different ecological zones and ways to minimize the impact of tourism activity is provided.

RESTRICTIONS: When tipping locals, avoid giving candy; pencils and notebooks better

REPRESENTATIVE MENU: *Breakfast*—fresh fruit, porridge; *Lunch*—fresh fruit, sandwiches; *Dinner*—vegetable soup, hot main course with meat, rice, potatoes, vegetables; *Alcohol*—wine with dinner

ABOUT THE COMPANY
TERRA ADVENTURES, INC.
70-15 Nansen St., Forest Hills, NY 11375, (718) 520-1845

Terra Adventures, Inc., is a new for-profit travel company with a staff of three focusing on Central and South America. It strives to promote consciousness on ecological endangerment of certain tourist areas. It works closely with locals to maintain Machu Picchu. It also gives money to Association for Protection and Conservation of Nature in Peru, as well as being an Ecology and Conservation Organization (ECO) group in Peru. Terra Adventures, Inc., practices low-impact camping.

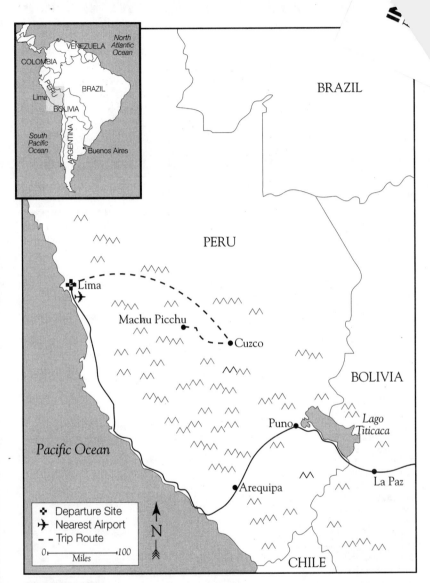

ECOFOCUS This area is heavily impacted by tourist traffic. The trash tourists leave behind is a danger to native wildlife.

∎TERNATIONAL RAINFOREST WORKSHOP
xplore the Amazon rainforest of Peru with International Expeditions

International Rainforest Workshops expose participants to something contagious in this region–a love and respect for rainforests which will spread unrestrained and infect those that share their experience. How? Because actually being in a rainforest is unforgettable. You walk down a pathway covered in vines and listen to the vast myriad of sounds and smells around you. Gazing high above your head, you catch a glimpse of a three-toed sloth. As you glide silently in a canoe down narrow waterways over-hung with chirping, rustling, squawking nests of green vibrating with life, you realize that no place on Earth can compare. You will have the opportu-nity to exchange ideas with others who share your interest in and concern about rainforests, as well as to hear from experts in the conservation field exactly what is going on and what you can do about it. You will participate in small group workshops involving hands-on field experience. This Inter-national Rainforest Expedition will provide you with valuable knowledge and insight to take home.

ABOUT THE TRIP

LENGTH OF TRIP: 8 days

DEPARTURE DATES: Mar.

TOTAL COST: $1,598; optional Machu Picchu extension $898

TERMS: Check; $300 deposit; refund of all but $25 for cancellations made in writing more than 60 days in advance

DEPARTURE: The trip begins once you arrive at the Miami airport.

AGE/FITNESS REQUIREMENTS: None

SPECIAL SKILLS NEEDED: Enthusiasm and a desire to learn

TYPE OF WEATHER: 80°- 85° during the day, 65° at night; evening rain

OPTIMAL CLOTHING: Hiking or walking shoes, cool cotton clothes, long-sleeved shirts

GEAR PROVIDED: Not necessary

GEAR REQUIRED: Old clothes, shoes with grip, backpack, binoculars, camera

CONCESSIONS: Film, personal items and sou-venirs are all available.

WHAT NOT TO BRING: Appliances, makeup, dressy clothes

YOU WILL ENCOUNTER: Many indigenous people; excellent opportunity to view birds, plants and insects

SPECIAL OPPORTUNITIES: Helping researcher, activities with local people

ECO-INTERPRETATION: Leaders have background in conservation, skilled local guides are used and partici-pants take part in an incredible hands-on workshop.

RESTRICTIONS: Don't flaunt expensive jewelry or belongings.

REPRESENTATIVE MENU: *Breakfast*—fruit, juice, eggs, toast, rice; *Lunch*—rice, beans, meat, fruit, vegetables; *Dinner*—meat or fish, tropical fruit, choice of vegetables, dessert; *Alcohol*—fun bar with singing, dancing; local beer and rum

ABOUT THE COMPANY

INTERNATIONAL EXPEDITIONS
One Environs Park, Helena, AL 35080, (800) 633-4734
International Expeditions is a for-profit company, owned by Dr. Richard Ryel, that has been in business 11 years. It strives to stimulate understanding and appreciation of the Earth's natu-ral wonders. An important outcome of its First International Rainforest Workshop was a $60,000 contribution to the establishment of the nonprofit Amazon Center for Environmen-tal Education and Research (ACEER). This organization will serve as a research base for sci-entists and provide an educational opportunity for ecotourists who visit the site.

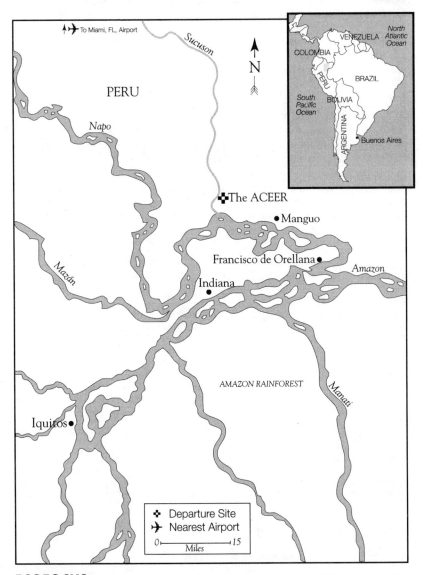

Peru

To Miami, FL, Airport

Sucuson

N

PERU

Napo

North Atlantic Ocean

VENEZUELA

COLOMBIA

BRAZIL

PERU

South Pacific Ocean

BOLIVIA

ARGENTINA

Buenos Aires

✚ The ACEER

● Manguo

Francisco de Orellana ●

Mazán

Indiana ●

Amazon

AMAZON RAINFOREST

Manati

Iquitos ●

✚ Departure Site
✈ Nearest Airport

0 ⊢——————⊣ 15
Miles

ECOFOCUS Logging and subsistence farming are a major source of deforestation in the Amazon rainforest of Peru.

PERUVIAN AMAZON DISCOVERY
Explore the Reserva Tahuayo in Peru with Amazonia Expeditions

This is the only commercial Amazon trip available in which the owners are highly qualified jungle guides. Truly fascinating and widely experienced, Paul Beaver, Ph.D., and Milly Sangama authored the book *Tales of the Peruvian Amazon*. On this expedition, you will travel over 100 miles upriver of Iquitos to find the best untouched wilderness available, including an exploration of the Reserva Tahuayo, the only place in the Amazon still home to red vakari monkeys. Operating from a base camp, you are offered as many as six itineraries each day. The native staff makes sure that each individual can enjoy a customized itinerary. If you're looking for an easygoing vacation with excellent food, this is for you. On the other hand, if rugged camping and wilderness survival sound more exciting, a true adventure may also be arranged. A typical day of this trip might include rising early for birdwatching, breakfast with the group and a hike through the jungle to see electric eels and lily pads. After lunch, you might hop in a boat and head out to watch pink dolphins play in the river waters. Supper is a time to relax, enjoy delicious food and prepare for a midnight outing along the river banks to look for caiman crocodiles.

ABOUT THE TRIP

LENGTH OF TRIP: 14 days

DEPARTURE DATES: At least once a month

TOTAL COST: $1,375

TERMS: Payment by check; $200 deposit; must cancel 48 hours prior to departure

DEPARTURE: Nearest airport is in Iquitos; guests depart by boat from Iquitos

AGE/FITNESS REQUIREMENTS: General good health

SPECIAL SKILLS NEEDED: None

TYPE OF WEATHER: Rainy season Nov.-May; temperatures range from the 70s to the 80s.

OPTIMAL CLOTHING: Cotton clothes

GEAR PROVIDED: Camping and jungle gear

GEAR REQUIRED: Backpack, flashlight, canteen

CONCESSIONS: Film, personal items and souvenirs can be purchased in Iquitos.

WHAT NOT TO BRING: Electrical appliances

YOU WILL ENCOUNTER: Indigenous people; wildlife, including macaws, marmosets, pink dolphins and electric eels

SPECIAL OPPORTUNITIES: Birdwatching, jungle survival instruction, photography, bass-type fishing for tucunare, deep-sea-type fishing for zungaro

ECO-INTERPRETATION: The head guide, Paul Beaver, has a Ph.D. in Biology and offers informative interpretation.

RESTRICTIONS: Careful consideration of impact on jungle

REPRESENTATIVE MENU: *Breakfast*—fruit, French toast, cereal; *Lunch*—2-3 salads, fish, vegetables; *Dinner*—pasta, 2-3 salads, potatoes, chicken, beans; *Alcohol*—beer sold, champagne for fiesta provided

ABOUT THE COMPANY

AMAZONIA EXPEDITIONS
1824 NW 102nd Wy., Gainesville, FL 32606, (800) 262-9669
A for-profit adventure-travel company specializing in trips to the Amazon area, Amazonia Expeditions has been in business 11 years and is corporately owned. It has made an agreement with a local community called Jaldar in which it helps with the town's medical facilities and school supplies and, in turn, the locals assist in developing a "reproductive reserve." This Moringo Reserve boasts a good reproductive habitat and the largest variety of flora and fauna of any ecosystem in the area. The majority of the 23-person staff is native South American.

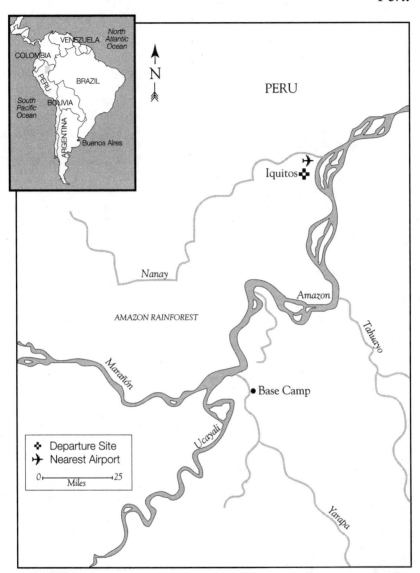

ECOFOCUS This area consists of a number of wilderness preserves, each with a different status. Pacaya-Samiria National Park is being considered for oil exploration by US companies. The exploration would take place in the green paradise area of the far-western lowland forests of the Amazon, which include a rich diversity of plant and animal life. Policies regulating this exploration have been shelved by the president of Peru, but could be activated anytime. It is essential that this area be preserved and protected from the impending threat of exploration and development.

ORINOCO ADVENTURE
Explore the Orinoco River in Venezuela with Amazon Explorers, Inc.

Otherwise known as "The Lost World," Venezuela is one of the most diverse and unknown destinations of South America. The Orinoco Adventure begins with a six-day dugout canoe trip down the mysterious and mighty Orinoco River. You will see colorful macaws and parrots, monkeys, caimans, anacondas and freshwater dolphins as your dugout drifts through the waters. You will have the opportunity to visit local Yabarana, Yekuana and Piaroa communities. On Day 8 you will navigate the Atabune Swamp and spend the night at the Tama-Tama Mission. Day 10 will take you on to the Yutaje Floodplains, from where you will fly over the strange, enigmatic Cerro Autana table mountain. Continuing the next day, you'll see Canaima, land of a thousand waterfalls. On Day 13, you will fly through the spray over magnificent Angel Falls, the highest in the world. These falls tumble over 3,000 feet, from the immense 200-square-mile top of the Auyantepui table mountain into the dense jungle below. That night you will camp in Canaima, before spending your final day wandering the streets of Caracas and dining at a typical Spanish restaurant before your departure the following day.

ABOUT THE TRIP
LENGTH OF TRIP: 12 days

DEPARTURE DATES: Every week

TOTAL COST: $1,998 including local airfares

TERMS: Checks; $150 deposit, full payment at least 45 days before departure; cancellations 46 days before trip, $25 handling fee; 45-31 days before, 75% refund, 30-16 days before, 50% refund, 15-7 days before, 25% refund, 6-0 days before, no refund

DEPARTURE: Arrive at Caracas Airport; trip begins there.

AGE/FITNESS REQUIREMENTS: All ages; good physical condition

SPECIAL SKILLS NEEDED: None

TYPE OF WEATHER: Warm to oppressively hot; rain possible; average temperature 78°

OPTIMAL CLOTHING: Lightweight shirts, both long and short sleeves, lightweight pants and shorts, swimsuit, sneakers, rain poncho; anti-malaria pills recommended

GEAR PROVIDED: Group equipment

GEAR REQUIRED: Sunscreen, insect repellent, flashlight, hat, backpack

CONCESSIONS: Souvenirs and personal items can be purchased in Caracas; bring your own film.

WHAT NOT TO BRING: Rifles, side arms, bows, valuables

YOU WILL ENCOUNTER: Indigenous peoples; spectacular wildlife

SPECIAL OPPORTUNITIES: Fantastic photography

ECO-INTERPRETATION: Travelers learn about the condition of the river, its surroundings and the people living nearby.

RESTRICTIONS: No money exchanging with indigenous peoples

MEALS: In restaurants in Caracas; open-fire cooking on excursion

ABOUT THE COMPANY
AMAZON EXPLORERS, INC.
499 Ernston Rd., Parlin, NJ 08859, (800) 631-5650
In business 35 years, Amazon Explorers, Inc., was founded by Richard G. Brill and served as the first travel outfitter to take visitors down many South American rivers. It also was the first corporation to protect indigenous people from losing their land, investing money to help locals buy jungle plots. In addition, Amazon Explorers protects animals by not allowing rifles or hunting on their safaris.

ECOFOCUS The Amazon jungle is being steadily whittled away. A million trees a day fall in South America. Indigenous people are displaced as the jungle disappears, and their ways of life are irretrievably lost.

SOJOURN TO ANTARCTICA
View Antarctica's wildlife with Mountain Travel/Sobek

Upon your arrival in Ushuaia, Argentina, the southernmost town in the world, you will embark on the *Professor Molchanov* down the Beagle Channel, lined with glaciers, forests of beech trees and the Darwin Cordillera. During the voyage you will be accompanied by albatross and giant petrels as you sail south from Argentina for 600 miles to Antarctica. The *Professor Molchanov* will take you along the Danco Coast to Paradise Harbor and into the narrow passageway of the Neumeyer Channel, where you will encounter icebergs calving, whales and penguins swimming next to the ship, seals floating on ice and a variety of birds. You will visit Deception Island, with one of the only active volcanoes on the continent. The cruise will visit the research stations of several different countries and travel to Half Moon Island, Port Lockroy, Caverville Island, Charlotte Bay, King George Island and Waterboat Point. The exploration of this pristine wilderness takes place in the company of scientists and expert naturalists during the austral summer, when the days are long and the temperatures are mild. Each day you will set out with guides in a Zodiac landing craft to cruise among ice floes and penguin rookeries. The return trip will take you north to the protected channels of Tierra del Fuego before you return to Ushuaia and fly home.

ABOUT THE TRIP

LENGTH OF TRIP: 17 days

DEPARTURE DATES: Nov. 16, 29, Dec. 12, 25, Jan. 7, 20, Feb. 2, 15

TOTAL COST: $4,500-$6,500

TERMS: MC, VISA, AmEx, check; deposit $1,000; cancellation fee of $500 with additional charges depending on time of notification

DEPARTURE: Arrive in Ushuaia and join the group on the *Professor Molchanov* to begin sailing through the Beagle Channel.

AGE/FITNESS REQUIREMENTS: General good health

SPECIAL SKILLS NEEDED: None

TYPE OF WEATHER: Cold, Antarctic climate; 20°-50°

OPTIMAL CLOTHING: Cold-weather clothing

GEAR PROVIDED: Parka

GEAR REQUIRED: Rubber boots, personal clothing, film and camera

CONCESSIONS: Film, souvenirs and personal items available on board

WHAT NOT TO BRING: Skis

YOU WILL ENCOUNTER: Researchers and scientists stationed in Antarctica; unique Antarctic wildlife.

SPECIAL OPPORTUNITIES: Hiking, boating and birdwatching

ECO-INTERPRETATION: Professional lectures are given on wildlife.

RESTRICTIONS: Avoid walking on mosses and lichens; maintain a respectful distance from the wildlife; do not remove anything from the environment.

REPRESENTATIVE MENU: *Breakfast*—omelettes, bread, bacon, bouillon; *Lunch*—buffet of international food; *Dinner*—steak, filets, pasta; *Alcohol*—can be purchased

ABOUT THE COMPANY

MOUNTAIN TRAVEL/SOBEK
6420 Fairmount Ave., El Cerrito, CA 94530, (800) 227-2384
Mountain Travel/Sobek donates money from trip proceeds to local organizations and uses local guides and lodgings. It specializes in adventure travel all over the world. Mountain Travel/Sobek is managed by Dick McGowan, head of the board of trustees, who combined the programs of Mountain Travel and Sobek in May 1991 and compiled a staff of 35.

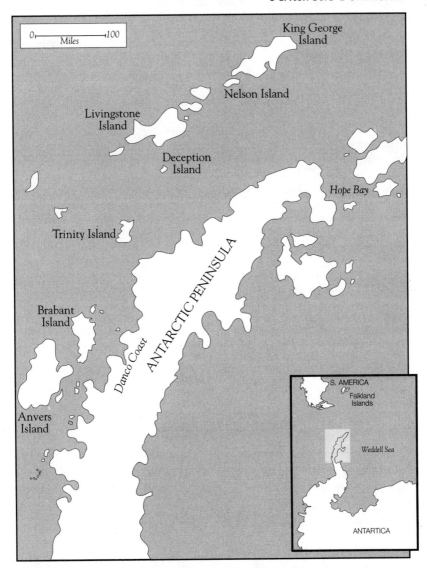

ECOFOCUS The hole in the ozone layer is affecting the stability of the Antarctica ecosystem, causing problems with exposure for the wildlife and the habitat. In addition, Antarctica is threatened with industrial exploitation in the form of proposed oil drilling and strip-mining. Counter proposals to turn Antarctica into a "World Park" could offer a permanent solution, and responsible tourism can help to pave the way toward saving this vast wilderness.

TUAREG CAMEL TREK
Trek the Sahara Desert in Algeria with Adventure Center

You peek out the window and see that your plane is landing on an airstrip in the middle of a sandy nowhere. Welcome to the Djanet Oasis. You hop aboard a rugged vehicle for the drive to Iharadj Oasis, where you join local guides and a team of camels. Referred to as "nobles of the desert," the tall Tuareg people are known for their striking, traditional blue robes. Their families have lived in this rugged country for thousands of years, proving themselves the best guides available. On this trek, your days begin with a breakfast prepared just for you. After packing your camera and canteen, you take off hiking over the dunes. Lunch comes early in the afternoon, followed by time to explore the local village or just relax. Continue hiking or traveling by camel into the evening, then set up camp or settle into local accommodations. Savor an evening meal, then snuggle down for the night. During this voyage, you will cross the arid wadis (dry river valleys) of Tekkat N'Tenere and bivouac in In Aramas, a natural shelter formed by erosion. You will also have the opportunity to explore a bizarre natural labyrinth leading to Agzel rock guelta, a natural pool enclosed by cliffs. With the Tuareg leading the way, your adventures through the Sahara are endless.

ABOUT THE TRIP

LENGTH OF TRIP: 10 days

DEPARTURE DATES: 1992: Mar. 20, Oct. 16, Nov. 13, Dec. 11; 1993: Jan. 15, Feb. 5, Mar. 26

TOTAL COST: $1,530-$1,610 (plus $70 local payment)

TERMS: Checks; $200 deposit; balance due 70 days prior

DEPARTURE: Arrive at Algiers airport; further travel provided

AGE/FITNESS REQUIREMENTS: At least 14 years old; good health and fitness

SPECIAL SKILLS NEEDED: None

TYPE OF WEATHER: Moderate temperatures ranging from the 50s to the 80s

OPTIMAL CLOTHING: Lightweight for days, warm for nights

GEAR PROVIDED: All items essential to trip

GEAR REQUIRED: Lightweight hiking boots, 4-season sleeping bag, mat

CONCESSIONS: Personal items and film should be purchased in advance.

WHAT NOT TO BRING: Olive-green or camouflage clothing that could be construed as military

YOU WILL ENCOUNTER: Tuareg people

SPECIAL OPPORTUNITIES: Spectacular photography of desert landscapes, Tassili paintings

ECO-INTERPRETATION: Tuareg guides share local knowledge.

RESTRICTIONS: Algeria is a Muslim country; dress conservatively and ask guide about etiquette when photographing people.

REPRESENTATIVE MENU: *Breakfast*—cereal, eggs, local fruit; *Lunch*—local fruits, tinned meats; *Dinner*—breads, meats, local fruits prepared at campsite; *Alcohol*—availability restricted by Muslim customs

ABOUT THE COMPANY

ADVENTURE CENTER
1311 63rd St., Ste. 200, Emeryville, CA 94608, (510) 654-1879, (800) 227-8747
For the past 13 years, Adventure Center has been offering exploratory holidays around the world for small groups who wish to rediscover the fragility and beauty of our world while making a minimal impact upon it. Local people are employed to operate tours, and contributions are made to the Earth Island Institute and East Africa Wildlife Society. Adventure Center is a for-profit organization owned by Johno Wells, with a staff of 12, striving for a more environmentally sustainable alternative to conventional travel.

Algeria

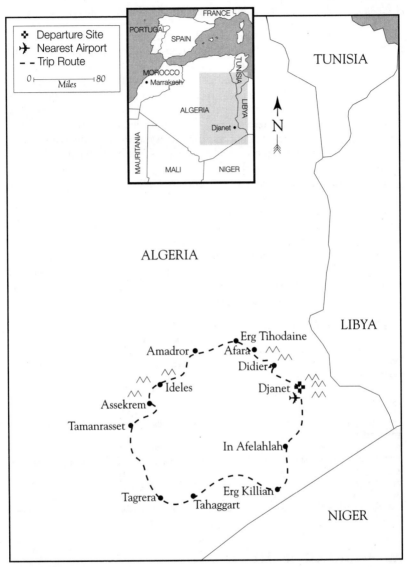

ECOFOCUS Algeria is plagued by localized environmental problems such as erosion and loss of arable land due to overgrazing.

ELEPHANT-BACK SAFARI
Ride elephant-back through Botswana with Ker and Downey Safaris

No, this isn't Indiana Jones. It's the only adventure travel operation available where African elephants are used to transport paying clients, and Randall Moore is the man to thank for it. Author of the book *Back to Africa*, Moore has initiated a relocation program for African elephants orphaned by "culling" operations in South Africa's Kruger National Park and adult elephants from zoos and other institutions in North America. The elephants used on these safaris—Abu, Bennie and Kathy—are examples of the animals who have successfully been returned to their home continent. Meandering at their own elephant pace, they will carry you where no one but wild game has trod and expose you to the unspoiled wonderland of the Okavango Delta in northern Botswana. With baby elephants in tow, you will view wildebeests, zebras, lions, giraffes, hippos and countless other magnificent creatures from about eight feet above the ground. Meals are fresh and delicious and cocktails are served nightly at base camp. Accommodations are luxurious East African-style safari tents positioned under majestic shade trees and serviced in five-star fashion. For the ultimate in extravagance and adventure, climb aboard an elephant with Ker and Downey Safaris and explore Africa in regal style while helping to support the preservation of these peaceful pachyderms.

ABOUT THE TRIP

LENGTH OF TRIP: 6 days

DEPARTURE DATES: Wednesdays from Mar. through Oct.

TOTAL COST: $3,425-$5,000 depending on size of group

TERMS: Cash, check, bank transfers; 30% deposit; cancellation fees are variable according to amount of advance notice

DEPARTURE: Arrive at airport in Maun; private air transport to Abu's camp provided

AGE/FITNESS REQUIREMENTS: None

SPECIAL SKILLS NEEDED: None

TYPE OF WEATHER: Cool evenings, warm to hot days, minimal rain; high 90°, low 45°

OPTIMAL CLOTHING: Casual, lightweight clothing in neutral colors; heavy sweaters and down jackets for evenings, comfortable walking shoes

GEAR PROVIDED: All group gear

GEAR REQUIRED: 12-volt adapters, camera

CONCESSIONS: Film and personal items available at different campsites

WHAT NOT TO BRING: Too much clothing

YOU WILL ENCOUNTER: Indigenous people on staff; unlimited wildlife viewing

SPECIAL OPPORTUNITIES: Mokoro (dug-out canoe) trips, night game drives

ECO-INTERPRETATION: Interpretation provided by qualified naturalist, environmentalist, conservationist and historian guides

RESTRICTIONS: Guides will discuss local traditions.

MEALS: Full meals of "bush cuisine," including fresh bread baked daily; special food requirements can be arranged

ABOUT THE COMPANY

KER AND DOWNEY SAFARIS
13201 Northwest Freeway, Ste. 850, Houston, TX 77040, (800) 423-4236

For 25 years, Ker and Downey Safaris has been leading traditional safaris throughout Africa. A for-profit company, the message it wishes to communicate to its guests is that unless they make a personal commitment, what they're seeing won't last. It has donated services, vehicles and finances to a myriad of local projects over the years. All necessary resources for its trips are flown or trucked in and all nonbiodegradable waste is removed from campsites. Ker and Downey is privately owned.

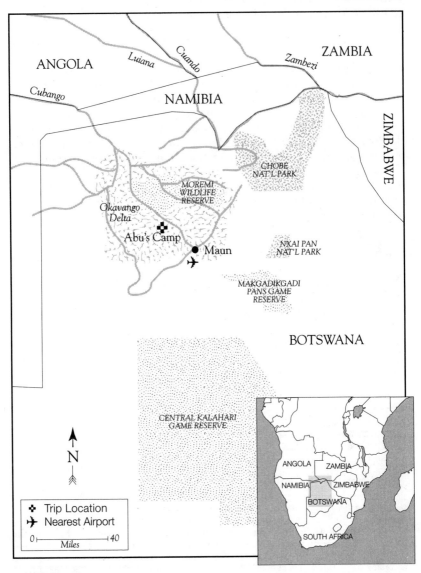

Botswana

ECOFOCUS The African elephant is endangered due to the ivory trade. Ten years ago, there were probably 1.5 million throughout Africa. Today, there are fewer than 500,000, and conservationists estimate that they are being killed at the rate of 80,000 a year. Botswana is being damaged by human encroachment and cattle ranching, and the Okavango Delta's water may soon be tapped to provide water for development. Botswana is home to many of the last remaining wild dogs in Africa.

HERBS AND AIDS
Research AIDS remedies with University Research Expeditions Program

As the threat of AIDS looms ever greater, with no definitive cure in sight, scientists are in a race against time. While less than 5 percent of the world's plants have been studied for their medicinal qualities, at least 50 or more plant species become extinct every day. Researchers are turning to traditional herbalists in the tropics to help identify those plants with potential pharmaceutical value. Since fewer and fewer people are being trained in traditional medicine, scientists are also struggling with the decline of indigenous knowledge of herbs. Join University of California scientists and others like yourself in gathering information from traditional Kenyan herbalists regarding the remedies they use. Travel throughout the wilds of eastern Africa with scientists collecting plant samples, viewing exotic wildlife and meeting native herbalists. The information you help gather may someday reveal a cure for AIDS.

ABOUT THE TRIP

LENGTH OF TRIP: Approximately 14 days

DEPARTURE DATES: July 28, Aug. 13, 28

TOTAL COST: $1,695

TERMS: Checks; $200 deposit; complete refund available until accepted to a program

DEPARTURE: The group will meet at Nairobi airport and further transportation will be provided.

AGE/FITNESS REQUIREMENTS: All ages; good health

SPECIAL SKILLS NEEDED: Interest in ethnobotany

TYPE OF WEATHER: Nairobi and Kitale can get cool at night. Mombasa, on the Indian Ocean, is hot and humid, with temperatures in the 80s and 90s.

OPTIMAL CLOTHING: Cotton clothes covering skin; boots are essential

GEAR PROVIDED: All technical gear

GEAR REQUIRED: Lightweight clothing, long-sleeved shirts, pants

CONCESSIONS: Film, personal items and souvenirs are available in Nairobi and Mombasa, but it is best to buy film in the US.

WHAT NOT TO BRING: Excessive clothing

YOU WILL ENCOUNTER: Local herbalists, incredible wildlife throughout Kenya

SPECIAL OPPORTUNITIES: Wonderful introduction to ethnobotany

ECO-INTERPRETATION: The project will be led by Moses Otsyula, a member of the Kenya Institute of Primate Research, and will focus on the use of native plants in medicinals.

RESTRICTIONS: Dress conservatively; show sensitivity to Kenyan culture.

MEALS: Local foods will be eaten at small restaurants.

ABOUT THE COMPANY

UNIVERSITY RESEARCH EXPEDITIONS PROGRAM
University of California, Berkeley, CA 94720, (510) 642-6586

Since 1976, UREP, a nonprofit organization, has been matching people from all walks of life with University of California scholars who are conducting scientific research around the globe. UREP supports the integration of indigenous knowledge with Western scientific studies, and involves researchers and students from host countries. Accommodations vary according to the trip, and donations will be a tax-deductible contribution to the University of California.

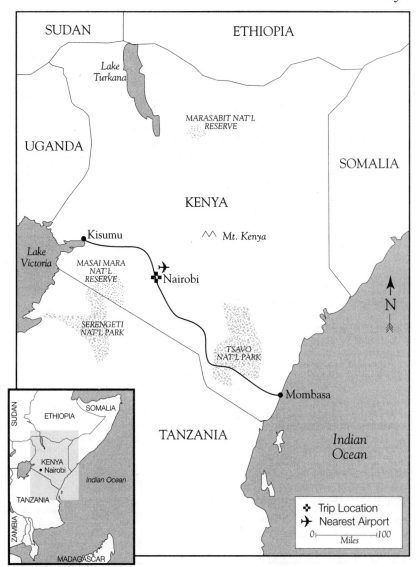

ECOFOCUS The rapid growth of Kenya's urban centers is destroying the environment and forcing herbalism into decline. Herbal remedies for all forms of incurable disease are crucial to medical research today.

KENYA WALKING SAFARI
Hike Kenya's wilderness with Taylor-Cassling, Ltd.

W ho wants to see Africa through the window of a van? Kenya Walking Safari by Taylor-Cassling, Ltd., takes you out of the vehicle and into the open air. Escape the crowds of tourists, step into the heart of the bush, observe magnificent wildlife and encounter fascinating tribal peoples. Hike game-filled bushlands in the old "Northern Frontier District," the wildest part of Kenya. View desert wildlife in Samburu Game Reserve and shadow elephants along the beautiful Uaso Nyiro River. Meet the "butterflies of the desert," the Samburu people, who wander with their herds of cattle and camels through the remote Mathews Range. Heading into the Great Rift Valley, marvel at millions of flamingos at Lake Nakuru National Park. Hike among old lava flows and bizarre rock formations in Hell's Gate Gorge. Stroll through the scenic Loita Hills on the Tanzania border, on an exciting cultural trek with the Masai. This is a rare opportunity for a cross-cultural exchange on a very personal level. Anticipate three days of rest and superb game viewing by Land Rover through Masai Mara Game Reserve, teeming with herds and huge prides of lions. Kenya is the most popular destination for safari in Africa because of its superior national parks and reserves. Here's an opportunity to avoid the crowds and enjoy an authentic adventure travel experience by foot.

ABOUT THE TRIP

LENGTH OF TRIP: 20 days
DEPARTURE DATES: June 7, July 5, Aug. 2, Sept. 6, Oct. 11, Dec. 20
TOTAL COST: $3,490 plus $320 park fees
TERMS: Cash, check, MC, VISA, AmEx; $500 deposit within 2 weeks of confirmation; $500 cancellation refund applied to another trip
DEPARTURE: Arrive in Nairobi airport; further transport provided
AGE/FITNESS REQUIREMENTS: All ages; good health
SPECIAL SKILLS NEEDED: None
TYPE OF WEATHER: Warm; 50°-85°
OPTIMAL CLOTHING: Lightweight hiking boots, hat, all-cotton clothing
GEAR PROVIDED: None
GEAR REQUIRED: Flashlight, clock, personal items

CONCESSIONS: Photos, souvenirs, personal items and maps available; purchase film in US
WHAT NOT TO BRING: Radios, bad attitude
YOU WILL ENCOUNTER: Indigenous peoples everywhere; endless wildlife viewing
SPECIAL OPPORTUNITIES: Extraordinary photography, some fishing, opportunity to balloon over animals
ECO-INTERPRETATION: Guides are graduates with an MA or Ph.D. in botany, conservation, natural history or ecology.
RESTRICTIONS: Be respectful of indigenous cultures.
REPRESENTATIVE MENU: *Breakfast*—canned fruit, local milk, eggs and cornmeal; *Lunch*—meat, cooked vegetables, tea, sodas; *Dinner*—pasta, meat, seafood, vegetables, fruit; *Alcohol*—beer available for purchase

ABOUT THE COMPANY

TAYLOR-CASSLING, LTD.
4880 Riverbend Rd., Boulder, CO 80301, (303) 442-8585
In business six years, Taylor-Cassling, Ltd., is a for-profit adventure travel outfitter run by a staff of four and owned by Albert Taylor and Gail Cassling, who have 40 years combined travel industry experience. It strives to promote general cultural awareness and preservation of our planet's wildlife and wilderness.Taylor-Cassling, Ltd., makes donations to Wildlife Conservancy in Kenya. Guides are both North American and local. All hiking occurs on established trails, all waste is packed out and only local items can be given to natives.

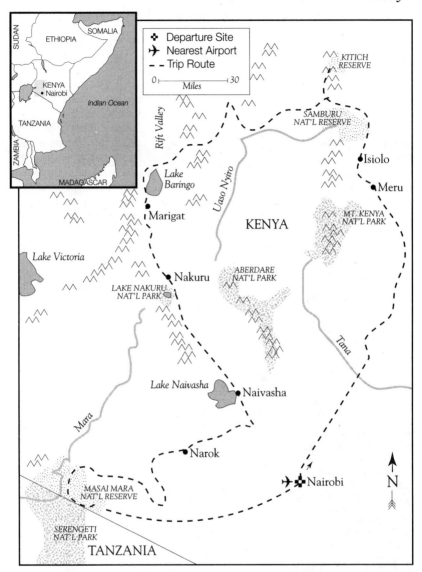

ECOFOCUS Due to lack of rain, droughts in Kenya are getting longer and longer. Rivers are depleted because of natural phenomena and increasing population.

MADAGASCAR DISCOVERY
Explore Madagascar with Above the Clouds Trekking

Don't ask for a detailed map of this expedition. The area is so remote that no one has even been there to map it! You will explore Antananarivo, Madagascar's charming capital city, walk through a dense forest viewing lemurs and other wildlife at Perinet, a nature reserve just to the east, and visit Sambava in the northeast for some cultural sightseeing. The heart of the trip begins in Andapa in the interior and includes a week-long circumambulation of the Marojezy massif, at 7,000 feet one of the tallest points on the island. The hiking days can be long, averaging 15 miles over sometimes-tricky terrain. However, the amazing scenery and the remote villages you will visit make this an extraordinary trip. The quintessential quote from an earlier trip in this area came from an old man in one of the villages halfway around the mountain: "There's a lot of you folks passing through here these days. Why, just last year there were two Frenchmen." After the trek, you'll investigate Montagnes d'Ambre National Park and Nosy Konba, the island where the black lemur is considered sacred. Take a trek where few foreigners have ever trod and gain insights about the danger of losing cultural and environmental resources in this unique, fragile area.

ABOUT THE TRIP

LENGTH OF TRIP: 25 days

DEPARTURE DATES: May 6, Aug. 26

TOTAL COST: $3,330

TERMS: VISA, MC, AmEx; $400 deposit; cancellations up to 60 days prior to departure.

DEPARTURE: Arrive at Antananarivo airport; company vehicle provided for transportation from airport to trailhead

AGE/FITNESS REQUIREMENTS: Any age; good health

SPECIAL SKILLS NEEDED: Stamina, patience

TYPE OF WEATHER: Warm days, cool nights in mountains; warm and pleasant on islands and coast. Temperatures in the upper 80s in the day and 40s to 50s at night.

OPTIMAL CLOTHING: Hiking boots, expeditionary sports sandals, layered clothing

GEAR PROVIDED: First-aid kit

GEAR REQUIRED: Trekking clothing, beach clothing, backpack, toiletries, mosquito net

CONCESSIONS: Film generally not available; some personal items and souvenirs available in Antananarivo shops and at the open market

YOU WILL ENCOUNTER: Indigenous Indonesians, Arabs and Africans; exotic wildlife including lemurs, chameleons and much more

SPECIAL OPPORTUNITIES: Snorkeling off Nosi-Bé and Nosy-Komba

ECO-INTERPRETATION: Natural history interpretation and cultural and historical background will be provided throughout by both American and local guides.

RESTRICTIONS: No tank tops, no shorts for women, no public affection

REPRESENTATIVE MENU: *Breakfast*—eggs, bread; *Lunch*—rice, spicy vegetables; *Dinner*—rice, spicy mea-and-vegetable stews, fried meats and fish, French cuisine available in hotels; *Alcohol*—some local beer and rum in cities

ABOUT THE COMPANY

ABOVE THE CLOUDS TREKKING
P.O. Box 398, Worcester, MA 01602-0398, (508) 799-4499

Above the Clouds Trekking strives to provide cross-cultural experiences for both its travelers and its hosts in remote, non-touristed areas. In business ten years, it is a for-profit organization run by a staff of three and owned by Steve Conlon. It specializes in culturally and environmentally sensitive remote-area treks in the Himalayas, Europe and Africa.

Madagascar

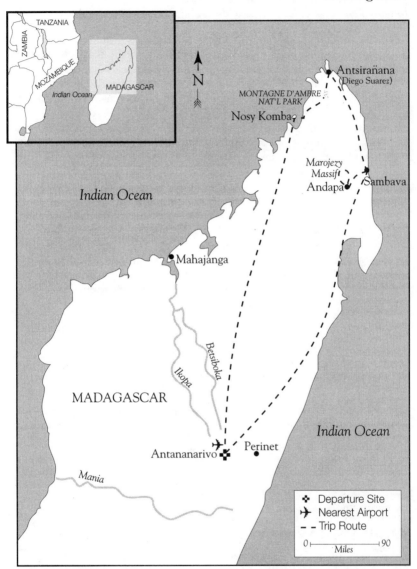

ECOFOCUS Deforestation is an acute concern in Madagascar due to population pressure. The resulting erosion is also devastating. Lemurs and other endemic species are therefore endangered, as are the indigenous cultures.

HANNIBAL'S ELEPHANTS
Research endangered elephants with Foundation for Field Research

A herd of around 300 Gourma elephants constitute the species' northernmost population and the last still existent in the Sahel. Neighboring Mauritania's elephants are now presumed extinct. All of Senegal's 28 remaining elephants can be spotted while driving through Niokolo Koba National Park, and the elephants remaining in Niger, Chad and the Sudan are but a few scattered individuals. Although poaching does not pose a problem in Mali, there is severe water shortage. Participating in Hannibal's Elephants gives you an opportunity to learn more about this devastating situation while helping to make a difference. Join hands with Friends of Animals, the Malian government, actress Brigitte Bardot and the Foundation for Field Research in efforts to preserve these precious pachyderms. Work side by side with scientists, tracking, counting and identifying various families of elephants. Malian botanist Assetou Kanoute will conduct research on the Gourma elephants' diet, while a Malian sociologist will inquire into why the elephants are not hunted, even though ivory prices have soared. Investigation into how the local people and elephants can share water sources will also be studied. Friends of Animals is currently targeting the digging out of silted water holes with help from the foundation.

ABOUT THE TRIP

LENGTH OF TRIP: 14 days

DEPARTURE DATES: July 5, 19, Aug. 2, 30

TOTAL COST: $1,672

TERMS: Checks; $250 deposit; trip cost is tax-deductible and in the case of cancellation, deposit can be used toward other trips

DEPARTURE: Arrive at airport in Bamako; take taxi to local hotel.

AGE/FITNESS REQUIREMENTS: 14 years or older

SPECIAL SKILLS NEEDED: None

TYPE OF WEATHER: Hot and windy desert; temperatures in the 90s in the day and 80s at night

OPTIMAL CLOTHING: Light cotton clothes, good boots

GEAR PROVIDED: Tent, sleeping pad

GEAR REQUIRED: Sleeping bag

CONCESSIONS: Photos, personal items, souvenirs and maps available; bring your own film

WHAT NOT TO BRING: Electrical appliances, guns, drugs

YOU WILL ENCOUNTER: Indigenous peoples; excellent exposure to elephants

SPECIAL OPPORTUNITIES: Photography

ECO-INTERPRETATION: Volunteers assist wildlife biologists and receive training on the spot.

RESTRICTIONS: Sensitivity to customs of Islamic culture

REPRESENTATIVE MENU: *Breakfast*—pancakes, orange juice, coffee; *Lunch*—sandwiches, fruit; *Dinner*—local meats, chicken, vegetables, fruit; *Alcohol*—available at some villages

ABOUT THE COMPANY

FOUNDATION FOR FIELD RESEARCH
P.O. Box 2010, Alpine, CA 91903, (619) 445-9264

Foundation for Field Research is a nonprofit organization in business ten years with a staff of nine people. Their aim is to use money from tourism to research various environmental issues and contribute information to conservation efforts. This project is funded in cooperation with Friends of Animals.

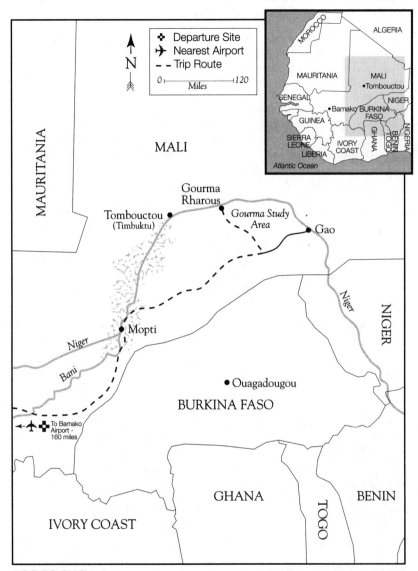

Mali

Departure Site
Nearest Airport
Trip Route

N

0 |—————————| 120
Miles

MALI

MAURITANIA

ALGERIA

MAURITANIA

MALI
•Tombouctou

SENEGAL

NIGER

•Bamako BURKINA
FASO

GUINEA

SIERRA
LEONE

IVORY
COAST

GHANA

BENIN
TOGO

NIGERIA

LIBERIA

Atlantic Ocean

Gourma
Rharous

Tombouctou
(Timbuktu)

*Gourma Study
Area*

Gao

Niger

NIGER

Niger

Mopti

Bani

•Ouagadougou

BURKINA FASO

To Bamako
Airport -
160 miles

GHANA

BENIN

TOGO

IVORY COAST

ECOFOCUS Less than 300 elephants remain in Mali due to conflicts between farmers and elephants over limited water resources. This small population of elephants is the last biologically viable group of elephants in this country.

NAMIBIA DESERT SAFARI
Explore Namibia with Wilderness Travel

After years of South African oppression, newly freed Namibia is starting to be recognized as the long-hidden jewel it is. Wilderness Travel's Namibia Desert Safari offers a thorough exploration of the fascinating and diverse natural and cultural environments of this newly independent country. Camp in the Namib Desert, exploring some of the highest sand dunes in the world (up to 1,000 feet in height). Discover the widest variety of landscapes in any national park in Africa at Namib-Nauklauft National Park. En route to Damaraland, investigate Twyfelfontein, an area ornamented with tribal rock art and an extraordinary array of wildlife, as well as the rugged and dangerous Skeleton Coast. The seldom-visited desert, Kaokoveld, exposes specialized flora and fauna, mirages and sand dunes and an opportunity to meet the nomadic Ovahimba people. Visit the magnificent Etosha National Park, where you can feast your eyes on lions, cheetahs, leopards, zebras, wildebeests, elands, rhinos, dik-diks and over 300 species of birds. And this is just the beginning. You will travel in a comfortable four-wheel-drive vehicle and spend nights alternately camping or staying in guest houses, rustic bungalows and hotels. Experience the transformation of Namibia, and behold an extraordinary world of adventures and surprises.

ABOUT THE TRIP

LENGTH OF TRIP: 22 days

DEPARTURE DATES: May 1, July 3, Aug. 14, Sept. 4

TOTAL COST: $3,450

TERMS: Credit card, checks; $400 deposit; total refund available minus a cancellation fee

DEPARTURE: Arrive at Windhoek airport; further transport provided

AGE/FITNESS REQUIREMENTS: Over 15 years old; good health

SPECIAL SKILLS NEEDED: None

TYPE OF WEATHER: Warm, dry desert days, cool coast days

OPTIMAL CLOTHING: Lightweight, easily washable

GEAR PROVIDED: Camping gear, vehicles, reference books, camp crew

GEAR REQUIRED: Clothing, camera gear, sense of humor and flexibility

CONCESSIONS: Film available on trip

WHAT NOT TO BRING: Too much gear; weight limit is 35 pounds in a duffel bag

YOU WILL ENCOUNTER: Remote indigenous tribes of Ovahimba and Herero; excellent wildlife-viewing possibilities

SPECIAL OPPORTUNITIES: Flights, walks, photography

ECO-INTERPRETATION: There will be a professional guide throughout the trip, reference books available on the vehicle and post-trip conservation literature.

RESTRICTIONS: Establish a rapport with Ovahimba people before taking pictures. Gift giving is not encouraged.

REPRESENTATIVE MENU: *Breakfast*—cereal, coffee, eggs, bread; *Lunch*—cold cuts, salads, drinks; *Dinner*—soup, meat, vegetables, dessert; *Alcohol*—complimentary drinks in camp, buy your own in lodge

ABOUT THE COMPANY

WILDERNESS TRAVEL
801 Allston Way, Berkeley, CA 94710, (510) 548-0420, (800) 368-2794

A for-profit company, Wilderness Travel has been leading expeditions for the past 14 years to remote destinations all over the world, allowing their travelers to enjoy the "good life" without sacrificing the spirit of adventure. All trips are limited to small groups of travelers and are led by expert naturalists. Wilderness Travel is owned by William Abbott.

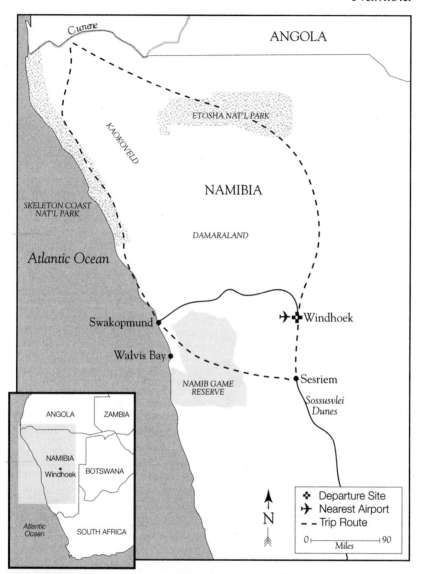

ECOFOCUS Namibia's national park wildlife preserves are experiencing pressures from both encroaching farmlands and wildlife poaching, particularly of elephants and rhinos. The mining complex of Tsumeb is expanding and could soon threaten the borders of nearby Etosha National Park. Also, open-cast mining is a hazard throughout the country. Namibia builds fences around its wildlife preserves which lessen the impact of poaching but interrupt the migratory patterns of animals.

WEST AFRICA PEOPLE-TO-PEOPLE
Cycle through Senegal and Gambia with International Bicycle Fund

See Senegal and Gambia in two weeks on two wheels. Those of you who appreciate the diversity of the world and don't mind the mild rigors of a bicycle tour, this trip is your ticket to satisfaction. While others measure their world tours by rolls of film exposed, miles traveled or entries stamped in their passports, you will go home with the doors of your mind and heart swung wide open. The first three days will be spent exploring Dakar and Goree Island. Next, you will cycle through the agricultural region of the western Sahel. Days 7 through 9 include visiting towns, villages and Banjul, in the estuary area of the Saloum and Gambia rivers. The final leg of this trip will offer you the lush beauty and cultural richness of the Casamance region. These 330 miles of cycling terrain are 90 percent paved and almost entirely flat, so you can focus on absorbing your surroundings. You will be guided by one Westerner with a background in area studies and one country national who is fluent in English and other appropriate local languages, as well as being an expert on West African history, culture and ecology. Enjoy incredible birdwatching, experience rural cultures and visit historic sites, all at a personal level and a human pace. Explore Senegal and Gambia with IBF and help promote bicycle transport, economic development and international understanding.

ABOUT THE TRIP
LENGTH OF TRIP: 14 days

DEPARTURE DATES: Oct. 15, Nov. 1

TOTAL COST: $1,090

TERMS: Checks; $300 deposit; $50 nonrefundable registration fee, $300 refund if cancellation between 60 and 30 days before departure.

DEPARTURE: Arrive in Dakar airport; transport provided to departure site

AGE/FITNESS REQUIREMENTS: All ages; moderate fitness

SPECIAL SKILLS NEEDED: Basic bicycle-riding skills

TYPE OF WEATHER: Cool part of dry season; sunny, warm to hot, not too humid

OPTIMAL CLOTHING: Cotton clothes; sturdy, multi-purpose footwear

GEAR PROVIDED: Rental bikes

GEAR REQUIRED: Clothing, first-aid kit, photo equipment

CONCESSIONS: Personal items, film and souvenirs may be purchased in Dakar, Banjul and Ziguinchor.

WHAT NOT TO BRING: Any unnecessary luggage

YOU WILL ENCOUNTER: Indigenous peoples; magnificent birdwatching possibilities

SPECIAL OPPORTUNITIES: Birdwatching, education, photography, walking

ECO-INTERPRETATION: Extensive environmental interpretation is provided. Guides are experts in the ecology, natural and social history, and native culture of the area.

RESTRICTIONS: Travel to meet, not conquer; listen and observe.

REPRESENTATIVE MENU: *Breakfast*—eggs, bread, coffee, tea; *Lunch*—fruit and baked goods; *Dinner*—excellent local cuisine; *Alcohol*—drinking discouraged until end of day; stay consistent with local standards regarding alcohol

ABOUT THE COMPANY
INTERNATIONAL BICYCLE FUND
4887-B Columbia Dr. S., Seattle, WA 98108, (206) 628-9314

In business ten years, the International Bicycle Fund is a nonprofit organization sponsoring bicycle trips, primarily in Africa, that assist economic development projects. IBF gives in-kind and financial donations to its host countries and provides training to locals. Ecotravel with IBF assures environmental friendliness, cultural sensitivity and benefits to the local economy.

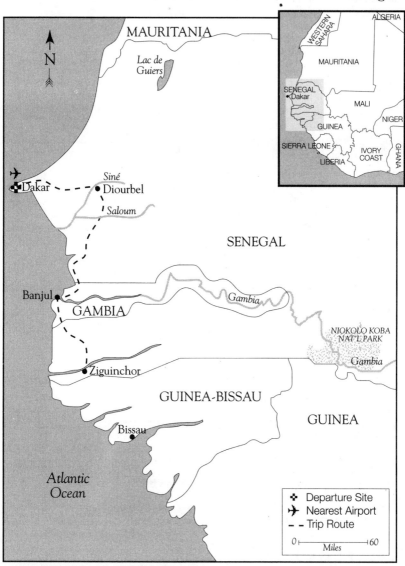

Senegal

ECOFOCUS This area of West Africa is plagued by deforestation. In urban areas, congestion, and air and water quality are concerns. The Wildlife trade is resulting in a rapid decline of species and Senegal is currently one of the top exporters of live parrots in the world.

SEYCHELLES ISLAND EXPLORER
Travel the Seychelles Islands of the Indian Ocean with Tamu Safaris

Inhabited by humans for barely 200 years, cast like shimmering jewels in an ocean that abounds in marine life, the Seychelles may well be the most beautiful islands in the world. They have been called "the Galapagos of the East," where rare birds, reptiles and plants have evolved to the unique rhythm of life on these isolated islands. On this special safari, you can explore the three main islands of the Seychelles: Mahé, Praslin and La Digue. Time will be taken to learn about their natural history, meet with local people and stay in small guest houses and hotels. On Cousin Island you can search for some of the world's 300 remaining brush warblers. Hiking on Praslin will take you into the Valle de Mai to learn about a palm forest that some scientists believe is a survivor of the pre-dinosaur era. La Digue offers the opportunity to snorkel amid some of the 200 species of fish and 150 types of coral found here. Hike among the giant granite boulders that are considered to be the remains of Gondwanaland, the continent that existed before Africa and India split apart. This special safari is for those travelers who wish to combine learning about natural history with the adventure of exploring one of the world's most remote and spectacularly beautiful island archipelagos.

ABOUT THE TRIP

LENGTH OF TRIP: 10 days (can be extended)

DEPARTURE DATES: June, July, Feb., Mar.

TOTAL COST: Varies

TERMS: Checks; $500 deposit; 25% cancellation fee between 60 and 30 days before departure

DEPARTURE: Arrive at the airport in Victoria; further transportation provided

AGE/FITNESS REQUIREMENTS: All ages; good health

SPECIAL SKILLS NEEDED: None; certification for optional scuba diving

TYPE OF WEATHER: Sunny with occasional rain during Feb. or Mar.

OPTIMAL CLOTHING: Lightweight summer clothing, sweater or jacket, light sneakers or sandals for reef walking

GEAR PROVIDED: Bicycles, scuba and snorkeling gear available for rent

GEAR REQUIRED: None

CONCESSIONS: Photos and maps available

WHAT NOT TO BRING: Excessive clothes; dress light and casual

YOU WILL ENCOUNTER: Local people; exceptional wildlife and marine life

SPECIAL OPPORTUNITIES: Tropical hiking, swimming, sailing, fishing, scuba diving and snorkeling

ECO-INTERPRETATION: Discussion about the formation of the islands, the rare palm forest of Praslin Island, endemic bird life, insects, reptiles, etc.

RESTRICTIONS: Must be courteous and understanding of the challenge of remote island living

REPRESENTATIVE MENU: *Breakfast*—continental; *Lunch*—fresh fish; *Dinner*—wonderful Creole cooking: seafoods, chicken, etc.; *Alcohol*—available in moderation

ABOUT THE COMPANY
TAMU SAFARIS
P.O. Box 247, West Chesterfield, NH 03466, (802) 257-2607, (800) 766-9199
Tamu Safaris is a for-profit company in business five years, with a staff of two full-time and two part-time employees. It specializes in private safaris and educational tours to six African countries, utilizing culturally sensitive, environmentally responsible methods of travel. "Small is beautiful" is the company's philosophy. Tamu Safaris supports local initiatives on conservation and is an active leader in developing and promoting ecotourism in Africa.

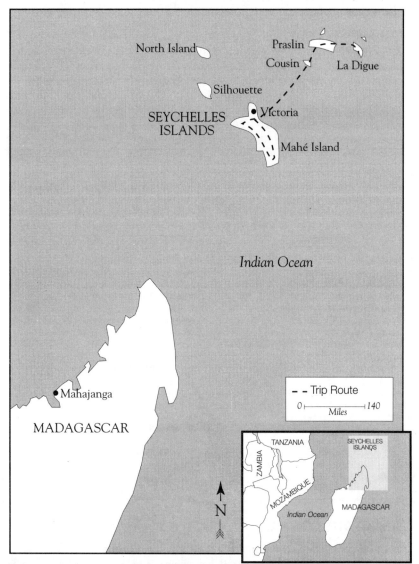

ECOFOCUS Fresh water is in short supply on these islands, making conservation of existing water supplies crucial. There are a large number of indigenous species unique to the Seychelles. The preservation of their natural habitat is crucial to their survival. Some species currently in danger include the hawksbill turtle, which is being poached, and the coco-de-mer coconut, which can only survive in a limited ecological system. The main force threatening this ecosystem is the influx of tourists. For this reason, the government only allows high-paying visitors in order to limit the numbers of people.

GORILLAS, CHIMPS AND WILDLIFE
View wildlife in and around Tanzania, with Overseas Adventure Travel

Picture yourself sitting in the Virunga Mountains within ten feet of an endangered gorilla group. Imagine tracking chimpanzees in the famed Gombe Stream Park, where Jane Goodall has studied chimpanzees for over 30 years. Now envision yourself hopping aboard a safari through Tanzania's classic parks. When you get sleepy, just stretch out in your tent under the wide-open African skies. Are you dreaming? Believe it or not, this dream can become a reality with Overseas Adventure Travel. Tanzania game parks you will visit include Serengeti, Ngorongoro Crater and Manyara, which offer an extraordinary array of exotic wildlife. You will be traveling in a specially designed vehicle large enough to carry 16 passengers. Your local guides are thoroughly trained by Overseas Adventure Travel. They are extremely knowledgeable and concerned about the preservation of wildlife and the natural environment. They are also expert game-spotters and know everything there is to know about species identification and animal behavior. This trip will also give you the opportunity to experience local culture while traveling through rural areas.

ABOUT THE TRIP

LENGTH OF TRIP: 27 days

DEPARTURE DATES: Jan. 9, Feb. 7, Mar. 5, July 9, Aug. 7, Sept. 3, Oct. 2, Dec. 11

TOTAL COST: $3,725

TERMS: Major credit cards; $500 nonrefundable deposit

DEPARTURE: Arrive at Kilimanjaro airport; further transport provided

AGE/FITNESS REQUIREMENTS: Good health; medical form required

SPECIAL SKILLS NEEDED: None

TYPE OF WEATHER: Varies with the season; temperatures in the 50s-90s, rain in forests

OPTIMAL CLOTHING: Lightweight, waterproof hiking boots, layers, raingear

GEAR PROVIDED: Tents, sleeping pads, cooking gear, all group equipment

GEAR REQUIRED: Sleeping bag, personal items

CONCESSIONS: Some souvenirs available; purchase film in advance

WHAT NOT TO BRING: Brightly colored raingear

YOU WILL ENCOUNTER: Rural Tanzanians; excellent wildlife viewing including mountain gorillas and chimpanzees

SPECIAL OPPORTUNITIES: Stupendous photo opportunities, especially of gorillas

ECO-INTERPRETATION: Concerned and knowledgeable guides give explanations and presentations on the environment.

RESTRICTIONS: Muslim culture; no photos of natives, conservative dress

REPRESENTATIVE MENU: *Breakfast*—eggs, toast, etc.; *Lunch*—cold cuts, bread, salads; *Dinner*—soup, chicken, vegetables, potatoes, fruit salad, dessert; *Alcohol*—beer can be purchased by passengers

ABOUT THE COMPANY

OVERSEAS ADVENTURE TRAVEL
349 Broadway, Cambridge, MA 02139, (617) 876-0533, (800) 221-0814

Overseas Adventure Travel is a 15-year-old for-profit travel organization owned by Judi Wineland, specializing in travel to East Africa, the Americas, the Himalayas and the Pacific. Its staff of 12 strives to provide quality trips with excellent services to demonstrate a clear concern for the environment. They employ local guides and stay in locally owned hotels. For each guest, Overseas Adventure Travel donates $25 to the African Wildlife Federation.

ECOFOCUS The mountain gorillas of Rwanda are threatened by the war with Uganda, which encourages poaching and harms the tourist industry. At Parc des Volcans, gorillas occasionally get caught in traps set for other animals. Gorillas are also susceptible to human diseases. The chimpanzees of Gombe are endangered by local fishermen's wood-cutting and encroachment by local people on park land.

BLACK RHINO EXPEDITION
Track endangered black rhinos in Zimbabwe with Earthwatch

Just two decades ago, 65,000 black rhinos flourished throughout the southern African wilderness. Today, fewer than 4,000 remain. Over half of this population resides in Zimbabwe. Closet biologists and animal lovers, here is your opportunity to indulge in intensive work helping scientists collect data on a research expedition led by Dr. Sky Alibhai and Dr. Zoe Jewell of the University of London, experts on the natural history of Zimbabwe. You will arrive at the airport in Harare and hop aboard a charter flight headed for the vast Chirisa Safari Area, which will be your home for the next 10 days. You and your team will emerge at dawn from your camp in the bushveldt to seek out these rare and beautiful rhinos with an ecologist and an armed game scout. Meet researchers and scientists from around the world and actually join them in gathering crucial research on the tragically diminishing black rhino population. Help native researchers and scientists track rhinos, photograph them at close range for an identification bank, collect and test spoor samples and assist in developing film back at the station. Seize this unique opportunity to live among elephants, antelope, Cape buffalo and many more exotic predators in this remote and wild research site. More importantly, help preserve the magnificent biological diversity still existing on our planet.

ABOUT THE TRIP

LENGTH OF TRIP: Approximately 11 days

DEPARTURE DATES: July-Sept.

TOTAL COST: $2,200

TERMS: MC, VISA, AmEx, etc.; $250 deposit; cancellations accepted up to 90 days before departure.

DEPARTURE: Arrive at airport in Harare; air travel to remote research site is provided.

AGE/FITNESS REQUIREMENTS: Minimum 16 years old

SPECIAL SKILLS NEEDED: Photography (optional)

TYPE OF WEATHER: Hot and dry

OPTIMAL CLOTHING: Typical field clothes

GEAR PROVIDED: Community camping equipment

GEAR REQUIRED: Sleeping bag, flashlight, camera and film, personal items

CONCESSIONS: Film, personal items and souvenirs available in Harare

WHAT NOT TO BRING: White or bright-colored clothing, camouflage gear

YOU WILL ENCOUNTER: Extensive wildlife viewing, including rhinos, lions, leopards, hyenas and buffalo

SPECIAL OPPORTUNITIES: Abundant photography possibilities

ECO-INTERPRETATION: Dr. Sky Alibhai, an expert on natural history of Zimbabwe from the University of London, leads the expedition.

RESTRICTIONS: The black rhino is a dangerous animal; caution must be exercised.

REPRESENTATIVE MENU: *Breakfast*—cereal, toast, jam, etc.; *Lunch*—cheese, meat, bread, fresh fruit; *Dinner*—canned and dried foods, local diet; *Alcohol*—beer and wine may be bought in town

ABOUT THE COMPANY

EARTHWATCH
680 Mt. Auburn St., Box 403N, Watertown, MA 02272, (617) 926-8200
A tax-exempt, international institution founded in 1971, Earthwatch strives to document a changing world. Trips are guided by native researchers and scientists, and local aid is given through stipends for students and employees. Earthwatch is not politically affiliated and is devoted to research.

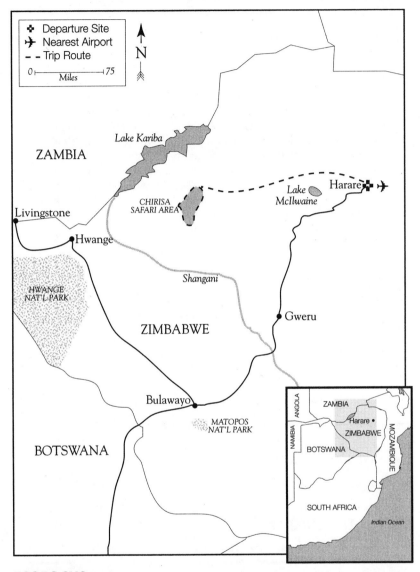

ECOFOCUS Black rhinos in Africa have been facing a severe population crisis for the past two decades. The 1970 population of 65,000 rhinos decreased to 3,392 rhinos by 1990. Zimbabwe and Namibia have managed to increase their numbers, and in a nonspecific survey done in 1990, Zimbabwe was the habitat for 1,700 rhinos, over one-half of the African rhino population. It is important to update the statistics on rhino populations in Zimbabwe to effectively influence park management and preservation policies, which will help the rhino population return to healthy numbers.

DAGESTAN: THE WILD CAUCASUS
Trek the Caucasus Mountains with InnerAsia Expeditions

When you arrive in Moscow, you will have two days to tour the city, visiting the Kremlin, Lenin's Mausoleum, St. Basil's Cathedral and a number of churches, convents and art galleries. Flying from Moscow to Makhachkala will take you to the Euro-Asian borderlands of the Caucasus Mountains, Europe's tallest range, home to 18,841-foot Mt. Elbruz. The next 10 days will be spent trekking at about 10,000 feet through the northern Caucasus, just north of the border of Azerbaijan, roughly along the Samur River. The route will take you through small villages and settlements far removed from traveled roads. Exploring in Dagestan, you will find thick forests, high snowy mountains, deep canyons and forts built by Alexander the Great. After the trek, you will return to the village of Jardan and drive east to the Caspian, where the group will board a boat to Makhachkala. You will have one more day to roam Moscow before your flight home. Dagestan is known for the 32 ethnic minorities that inhabit its valleys.

ABOUT THE TRIP

LENGTH OF TRIP: 13 days

DEPARTURE DATES: July 9

TOTAL COST: $2,550-$2,750

TERMS: MC, VISA, AmEx, cash, check; $500 deposit; cancellation fee of $300 increasing according to time of notification; no refund available 30 days before trip

DEPARTURE: Arrive in Moscow and join group at pre-arranged hotel.

AGE/FITNESS REQUIREMENTS: Very good physical condition

SPECIAL SKILLS NEEDED: Able to trek 5 hours each day with an average 2,000-foot elevation gain or loss per day

TYPE OF WEATHER: Day high 70° with some humidity; evenings 60°

OPTIMAL CLOTHING: Layers, comfortable hiking shoes

GEAR PROVIDED: All community camping equipment

GEAR REQUIRED: Raingear, duffel bag, sleeping bag, day pack, camera

CONCESSIONS: Bring all items with you; market availability in Moscow varies.

WHAT NOT TO BRING: Excess baggage

YOU WILL ENCOUNTER: Many cultural groups; native rhododendrons, sheep, bears

SPECIAL OPPORTUNITIES: Interaction with local village people, in an area that had been closed to visitors until two years ago, will allow you to witness a preserved culture.

ECO-INTERPRETATION: Guides will discuss cultural, environmental and natural history.

RESTRICTIONS: Cultural etiquette guidelines are provided. Waste-disposal, wood-use, water-treatment and campsite regulations must be obeyed.

REPRESENTATIVE MENU: *Breakfast*—hot cereal, biscuits, tea; *Lunch*—soup, bread, stew; *Dinner*—soup, Western and traditional local dishes; *Alcohol*—permitted but not provided

ABOUT THE COMPANY
INNERASIA EXPEDITIONS, INC.
2627 Lombard St., San Francisco, CA 94123, (415) 922-0448, (800) 777-8183
InnerAsia Expeditions is a for-profit company privately owned and organized by a staff of 12. Specializing in low-impact trips to Asia, Europe and the Americas, they support Woodlands Mountain Institute and Rodney Jackson's snow leopard research. This trip is contracted through a Moscow outfitter. InnerAsia Expeditions uses local guides and accommodations, educating and encouraging clients to contribute to environmental work upon their return from the trip.

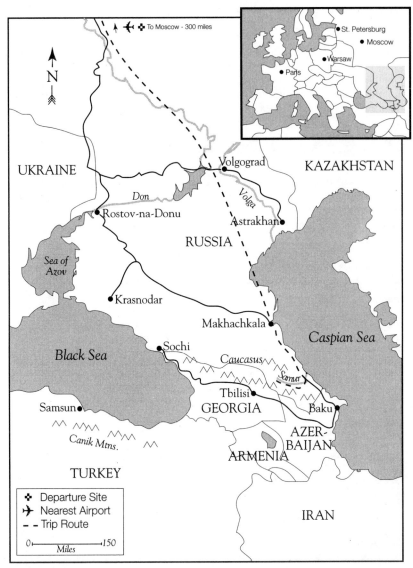

To Moscow - 300 miles

St. Petersburg

Moscow

Warsaw

Paris

UKRAINE

KAZAKHSTAN

Volgograd

Don

Volga

Rostov-na-Donu

Astrakhan

RUSSIA

Sea of
Azov

Krasnodar

Makhachkala

Caspian Sea

Black Sea

Sochi

Caucasus

Samur

Tbilisi

Baku

Samsun

GEORGIA

AZER-
BAIJAN

Canik Mtns.

ARMENIA

TURKEY

IRAN

✤ Departure Site
✈ Nearest Airport
- - Trip Route

0 ———————— 150
Miles

ECOFOCUS This land suffers from deforestation and overuse from recreation. In addition, oil refineries pollute the shore of the Caspian Sea.

EXPLORING EASTERN EUROPE
Hike, raft and climb with Mountain Travel/Sobek

Hike, raft and climb your way through Czechoslovakia and Poland. You will spend your first day sightseeing in Bratislava, then continue to Slovak Paradise National Park, where you will hike through gorges, caves, springs and waterfalls. Visit one of the largest castle ruins in Central Europe at Spissky Hrad and hike the next day to Sliezsky Dom, a mountain hotel near the base of Mt. Gerlach. After a night here, you will start hiking in the Slovak Tatras. The ambitious mountaineers in the group can climb Gerlach, the highest peak in Czechoslovakia. Those who choose to hike into Poland the next day will go over Mt. Polish Griebien and into the White Water Valley along the Polish-Czechoslovakian border. The options for the next few days include a walking tour of Zakopane, a mountain resort, visiting the nearby village of Chocholow, a hike to the Valley of Five Polish Lakes or through the western part of the Tatras. Pieniny National Park will be the focus of the next excursion, with hiking on a trail called Falcon Path. Spend the afternoon rafting through a scenic gorge on the Dunajec River. Before your last day in Warsaw, the cultural center of Poland, you will have a chance to visit Bialowiaska Nature Preserve, a primeval forest containing European bison, wolves, elks and numerous species of birds.

ABOUT THE TRIP

LENGTH OF TRIP: 15 days

DEPARTURE DATES: July, Aug., Sept.

TOTAL COST: $1,650

TERMS: MC, VISA, cash, check; deposit $400; cancellation $200 60 days prior to trip, but fees vary according to notification time

DEPARTURE: Arrive in Vienna; your guides will assist your travels to Bratislava.

AGE/FITNESS REQUIREMENTS: Good physical fitness with medical certificate

SPECIAL SKILLS NEEDED: Must be capable of hiking 8 hours in mountainous terrain

TYPE OF WEATHER: Mountain weather with temperatures of 45°-70°

OPTIMAL CLOTHING: Hiking clothes and boots, warm pants, waterproof/windproof parka

GEAR PROVIDED: Support vehicle

GEAR REQUIRED: Small day pack

CONCESSIONS: Varies according to market availability

YOU WILL ENCOUNTER: Bountiful wildlife, gorges, caves, springs and waterfalls

SPECIAL OPPORTUNITIES: Hiking, rafting and climbing

ECO-INTERPRETATION: Alternatives to damming the Danube River and national park preservation are discussed.

RESTRICTIONS: No hiking off trail

REPRESENTATIVE MENU: *Breakfast*—bread, coffee; *Lunch*—cheese, meat, bread; *Dinner*—traditional fare: meat, vegetables; *Alcohol*—available at local inns

ABOUT THE COMPANY

MOUNTAIN TRAVEL/SOBEK
6420 Fairmount Ave. El Cerrito, CA 94530-3606, (800) 227-2384
Mountain Travel and Sobek combined efforts in May 1991 and are presently Mountain Travel/Sobek, managed by Dick McGowan, head of the board of trustees. With a staff of 35, they specialize in adventure tours around the world. Mountain Travel/Sobek helps facilitate the preservation of national parks in Eastern Europe and employs local mountaineers and scientists as guides.

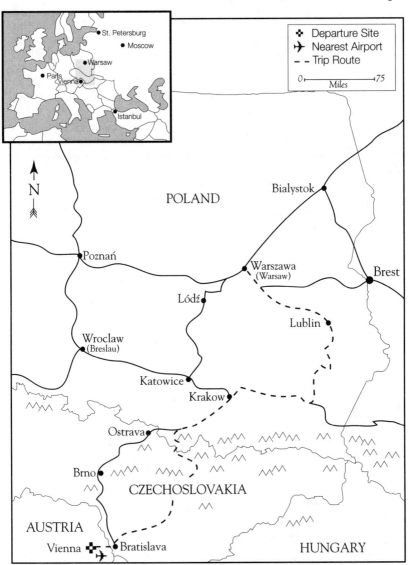

ECOFOCUS There are threats to the environment from industrial development and acid rain, but the most significant issue is the proposed dam on the Danube where Austria and Czechoslovakia meet. Preservation of this area is essential to the survival of an already threatened ecosystem and to the economic welfare of an area that could benefit if low-impact tourism succeeds.

SEA KAYAKING AND SCRAMBLING
Climb and kayak on the Isle of Skye with Outdoor Discoveries

Travel in Great Britain without being a typical tourist. Open yourself to the different culture and scenery of Scotland and England while exploring the outdoors. Sea kayak, camp, climb and enjoy your well-deserved rest in bed and breakfasts. This unforgettable adventure will take you from the sea to the mountains and through the countryside with British hospitality along the way. You will have 14 days to kayak around the Isle of Skye with opportunities for nontechnical rock scrambling in the Cuillin Hills. Try paddling on the Holme Pierrepont, an internationally used river-kayak slalom course, or try your luck fishing in the picturesque fishing village of Maillag, on the northwest coast of Scotland. Battle the elements of the Sound of Sleat and hope to find the sun-dappled moorlands and multi-colored sunsets behind the Cuillin Hills. Paddle past coastal forests, tumbling waterfalls, open moors and coastal hamlets, through glacier-carved lochs and by ancient castles and fortresses. Challenge yourself among majestic scenery while deepening your understanding of the local environmental issues.

ABOUT THE TRIP

LENGTH OF TRIP: 14 days

DEPARTURE DATES: July 26

TOTAL COST: $1,550

TERMS: Check, MC, VISA; 25% down 60 days in advance with a refund scale

DEPARTURE: Arrive at London's Gatwick Airport and drive in vans to Nottingham.

AGE/FITNESS REQUIREMENTS: Minimum of 16 years old with a parent, otherwise 18 years old

SPECIAL SKILLS NEEDED: Good conditioning

TYPE OF WEATHER: 65°-75° with either sun or rain

OPTIMAL CLOTHING: GoreTex, coated nylon raingear and soft-soled shoes

GEAR PROVIDED: Sea kayak, paddles, sprayskirts, life jackets, tents and cooking equipment

GEAR REQUIRED: Duffel bag, polypropylene, GoreTex, sleeping bag and pad

CONCESSIONS: In villages, film, personal items and souvenirs can usually be purchased.

WHAT NOT TO BRING: Alcohol, drugs, cellular phones or Walkmans

YOU WILL ENCOUNTER: Scenic coastlines and countryside, British culture

SPECIAL OPPORTUNITIES: Nontechnical climbing, contact with local people, flexible itinerary, fishing, and slalom kayaking

ECO-INTERPRETATION: Effects of deforestation on the Isle of Skye and the resulting erosion problems are discussed.

RESTRICTIONS: Come prepared to travel as the British do; be sensitive about British culture.

REPRESENTATIVE MENU: *Breakfast*—hot and cold cereal, pancakes; *Lunch*—cheese, meats, sandwiches, fruit; *Dinner*—hot stews, fresh vegetables, dessert; *Alcohol*—alcohol not recommended

ABOUT THE COMPANY
OUTDOOR DISCOVERIES
P.O. Box 7687, Tacoma, WA 98407, (206) 759-6555
Outdoor Discoveries is a for-profit company owned by Bob Stremba. It has been in business for three years specializing in educational adventures in Scotland, and the northwest and southwest US. The staff of five attempts to teach people about themselves, other cultures and the environment through minimum-impact back-country travel using local guides and lodgings.

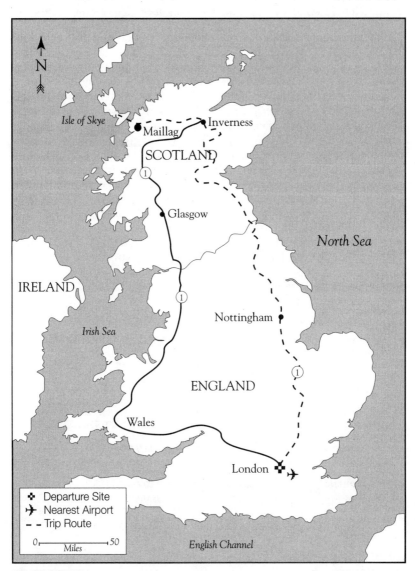

ECOFOCUS There are several environmental threats in Great Britain. The most significant is acid rain from European industry. Deforestation of almost all of Scotland's indigenous forests has resulted in erosion problems. The fisheries are depleted because of off-shore netting in northern England and there are high pollution levels in the Irish Sea.

NATURE TREK IN NORTHERN EUROPE
Birdwatch in Holland, Norway and Denmark with CAL Nature Tours

Enjoy the unique and abundant bird and marine life as well as the cultural richness of Norway, Holland and Denmark. Spend days exploring marshes and woodlands of the Dutch islands while watching the many bird species in breeding plumage. You will visit the North Sea Marine Mammal and Sea Bird Rehabilitation Center on Texel Island. After visiting the wetlands and woods of Lauwersmeer, return to civilization and visit historical Groningen and Amsterdam, where art and architecture merge to mold cities of charm and character. Enjoy the best of 91 countries at the Floriade Flower and Garden Show before you return to the wilderness and trek through the countryside surrounding Copenhagen. You will not be disappointed by plentiful birdwatching, treks through the woods and lakes of Utterslev Mose, and the beautiful Tivoli Gardens. Then visit the spectacular river valleys, glaciers and fjords of Norway. The grand finale of the trip is a special boat trip around Runde Island, the home of hundreds of thousands of nesting and feeding seabirds, sperm whales, gray seals and dolphins. The trip focuses on educational wildlife viewing while supporting local conservation groups and their attempts to maintain wetlands, protect breeding areas, maintain water quality and improve safety practices of oil transport in the North Sea.

ABOUT THE TRIP

LENGTH OF TRIP: 15 days
DEPARTURE DATES: May 1
TOTAL COST: $2,295
TERMS: Cash with a $200 deposit 60 days in advance; refund cost minus $50 up to 60 days in advance
DEPARTURE: Newark, NJ (price of trip includes airfare from Newark)
AGE/FITNESS REQUIREMENTS: None
SPECIAL SKILLS NEEDED: None
TYPE OF WEATHER: Mild, light spring rain with cool temperatures; 60°-75°
OPTIMAL CLOTHING: Jeans, sweatshirts and comfortable walking shoes
GEAR PROVIDED: Foul-weather gear
GEAR REQUIRED: Binoculars, bird identification book, comfortable, warm clothing, windbreaker

CONCESSIONS: Film, souvenirs and personal items can be purchased on trip
WHAT NOT TO BRING: Excess luggage or fancy clothing
YOU WILL ENCOUNTER: Indigenous people, fjords, birds, whales, reindeer, seals
SPECIAL OPPORTUNITIES: Endless photographic possibilities while nature trekking and visiting historical areas
ECO-INTERPRETATION: Habitat preservation and conservation of resources are discussed.
RESTRICTIONS: None
REPRESENTATIVE MENU: *Breakfast*—buffet, including bread, cheese, fish and meat; *Lunch*—herring, bread, cereals, fruit; *Dinner*—seafood, smoked lamb, Dutch pancakes; *Alcohol*—regional beer available

ABOUT THE COMPANY

CAL NATURE TOURS, INC.
7310 S.V.L. Box, Victorville, CA 92392, (619) 241-2322
CAL Nature Tours is a for-profit company that has been in business for seven years. Along with seven staff members, the owner, Thomas Gwin, runs trekking trips to Scandinavia, Holland, Eastern Europe, Jamaica and the southwestern US. Local naturalist guides and locally owned lodgings are used, and all trips promote habitat conservation as an alternative economic asset.

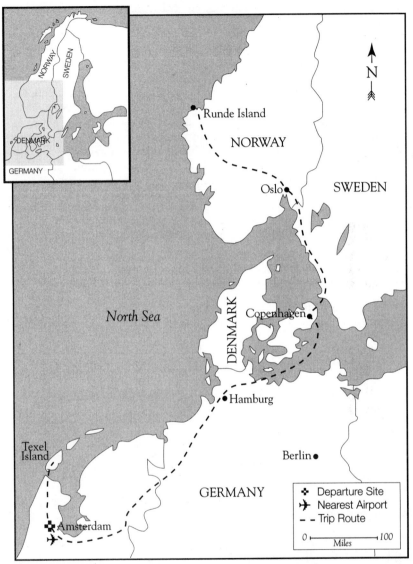

ECOFOCUS Northern Europe has a delicate ecosystem threatened by oil and toxic chemical waste dumping. The wildlife of these areas is endangered by pollutants in the water which are found in greater concentrations in animals higher on the food chain. Because these areas are breeding and nesting grounds for many species of birds, the effects of such pollution could be devastating.

ACROSS THE PYRENEES
Hike in France and Spain with Mountain Travel/Sobek

Begin your trip by soaking in the culture of Spain as you walk the narrow streets of San Sebastián. Warm up for the upcoming hike in the birch forest and shepherds' pastures of the Sierra de Sayoa. Shortly before daylight you will begin the trek, ascending through pine forests to alpine meadows inhabited by the Pyrenees mountain goat, the sarrios. Descending to the valley below, you will traverse along the Ara River and then climb to Mulets Pass, crossing the French boundary in full view of the Vignemale Cirque. Rest overnight at a traditional European-style mountain hut in Oulettes Refuge. By the next afternoon, the impressive limestone peaks will slip behind you as you leave the Gaube Valley. Pass the largest glacier in the Pyrenees before descending to the green pastures of Oulettes d'Ossoue Valley. If you seek a further challenge, climb Petit Vignemale. Stay overnight in the French village of Gavarnie and take in the view of the majestic amphitheater, where waterfalls pour off 3,000-foot walls. The next climb will bring you to Breche de Roland, an old smugglers' pass between France and Spain. Back in Spain, the path will descend through hanging glacier valleys to the national park of Ordesa. The grand finale is climbing Aneto, the highest peak of the Pyrenees.

ABOUT THE TRIP

LENGTH OF TRIP: 10 days

DEPARTURE DATES: May, June, Sept. and Oct.

TOTAL COST: $1,750

TERMS: Check, MC, VISA, AmEx, cash; deposit $400; cancellation fee starts at $200 and varies according to notification time

DEPARTURE: Arrive in Madrid or Barcelona and take a short flight to San Sebastián for an orientation with the tour leader.

AGE/FITNESS REQUIREMENTS: Must be a strong, experienced hiker able to hike 8-12 hours a day in rugged mountain terrain

SPECIAL SKILLS NEEDED: Hiking and scrambling experience

TYPE OF WEATHER: Warm and generally sunny, 40°-80°F, but varies according to time of year

OPTIMAL CLOTHING: Hiking gear

GEAR PROVIDED: None

GEAR REQUIRED: Large day pack

CONCESSIONS: Film, souvenirs and personal items are available in villages every evening.

YOU WILL ENCOUNTER: Indigenous people; local wildlife

SPECIAL OPPORTUNITIES: Swimming, climbing Petit Vignemale

ECO-INTERPRETATION: Guides will educate clients about nuclear power plants and development projects in the area.

RESTRICTIONS: None

REPRESENTATIVE MENU: *Breakfast*—baguettes, croissants, coffee; *Lunch*—bread, cheese, sausages; *Dinner*—Spanish fare: chicken, sausage, fish and vegetables; *Alcohol*—available

ABOUT THE COMPANY

MOUNTAIN TRAVEL/SOBEK
6420 Fairmount Ave., El Cerrito, CA 94530-3606, (800) 227-2384
Mountain Travel/Sobek donates money from trip proceeds to local organizations and uses local guides and lodgings. It specializes in adventure travel all over the world. Mountain Travel/Sobek is managed by Dick McGowan, head of the board of trustees, who combined the 20-year-old programs of Mountain Travel and Sobek in May 1991 to form a staff of 35.

Spain

Bay of Biscay

N

Bordeaux

FRANCE

Toulouse

Tarbes

San Sebastián

Pyrenees Mtns.

Perpignan

Gulf of Lions

Pamplona

Benasque

Zaragoza

Barcelona

SPAIN

Mediterranean Sea

St. Petersburg
Moscow
Warsaw
Paris
Vienna
Madrid
Istanbul

✤ Departure Site
✈ Nearest Airport
- - Trip Route

0 ⊢━━━━━┤ 50
Miles

ECOFOCUS A major highway through the Pyrenees, joining Pamplona and San Sebastián, is a threat to the historic farm country. Nuclear power plants also pose a threat to the environment. The trip leader, Martin Zabaleta, a famous Basque mountain climber, has been very active in both education about and protest of these potentially damaging projects.

JOURNEY TO THE EASTERN FRONTIER
Trek through Bhutan with Bhutan Travel Inc.

As a participant in this trip, you will be one of the lucky few to witness traditional life in this remote and untouched region, setting foot in a place where nature is respected and revered. You will explore Bhutan's far eastern frontier with a trek to the remote, high valleys of Mera Sakteng, home of the nomadic Drogpa people, combined with the spectacular annual festivals at Mongar and Tashigang. The Mera Sakteng valleys lie at an altitude of nearly 14,000 feet and are inhabited by an isolated group of yak-herding people. This part of Bhutan is far from the more popular routes, and offers a chance to witness traditions and ways of life that are centuries old. En route, see western, central and eastern Bhutan for a complete cross-country tour. At Bumdiling and Phobjika, you will go to the sanctuaries set aside for the extremely rare black-necked cranes, which migrate annually from Tibet and the Qinghai Plateau. The trip will be led by local guides and occasionally accompanied by US guides who are knowledgeable about environmental issues. The Bhutanese provide a model of conservative planning and sensitivity to the potential dangers of unregulated progress.

ABOUT THE TRIP

LENGTH OF TRIP: 22 days

DEPARTURE DATES: Nov. 28

TOTAL COST: $3,695

TERMS: Cash, check; $300 deposit; minimum cancellation fee $200

DEPARTURE: Passengers are met upon arrival in Bhutan; transport to departure site is provided in cars, minibuses and jeeps.

AGE/FITNESS REQUIREMENTS: Physical stamina and good general health

SPECIAL SKILLS NEEDED: None

TYPE OF WEATHER: Daytime temperatures around 50°, nighttime temperatures 30°-40° depending on altitude; generally sunny and dry

OPTIMAL CLOTHING: Layered, warm clothing, raingear, good hiking shoes

GEAR PROVIDED: Tents, cooking equipment, food, pack animals

GEAR REQUIRED: Sleeping bags, pads, day packs

CONCESSIONS: Local weavings can be purchased at the handicraft center in Thimphu; bring film and personal items from home.

WHAT NOT TO BRING: Ethnocentric or inflexible attitudes

YOU WILL ENCOUNTER: Nomadic Drogpa people; rare black-necked cranes, langur monkeys, many beautiful birds, butterflies, snow geese, wild yaks and possibly snow leopards

SPECIAL OPPORTUNITIES: Horseback riding

ECO-INTERPRETATION: This trip is led by a local Bhutanese guide and Steven LeClerq, naturalist and project worker in Nepal.

RESTRICTIONS: Entry into some temples restricted; no photography inside temples

REPRESENTATIVE MENU: *Breakfast*—juice, tea, coffee, eggs, bacon, toast, chapatis or rice; *Lunch*—hot lunch: soup, rice, noodles or potatoes, vegetables, meat, tea, coffee; trail lunch: sandwiches, cold potatoes, chapatis, fruit, juice; *Dinner*—soup, rice, noodles and/or potatoes, meat, vegetables, dessert, tea, coffee; *Alcohol*—available in hotels and shops

ABOUT THE COMPANY

BHUTAN TRAVEL INC.
120 E. 56th St., #1430, New York, NY 10022, (800) 950-9908
For 13 years, Bhutan Travel Inc. has been a for-profit company specializing in adventure, trekking and cultural exploration of Bhutan. They use local guides and accommodations and practice low-impact camping.

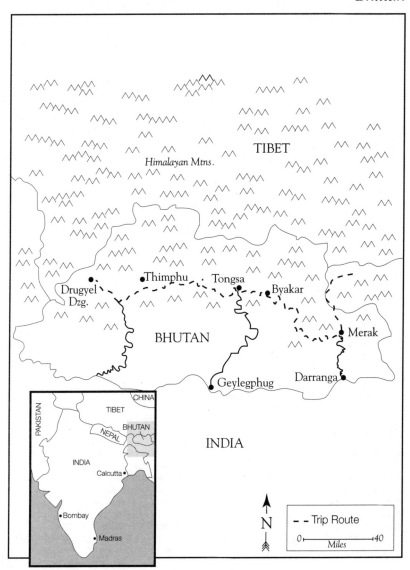

ECOFOCUS Bhutan is extremely careful about preserving its environment and regulating the rate of development. A government project in eastern Bhutan studies the effects of yaks on grasslands to prevent overgrazing. Deforestation is being controlled by laws that prohibit logging without permission. The World Wildlife Fund has developed a black-necked crane sanctuary and is working on the Manas preserve project and the King Jigme Dorji preserve.

MOUNTAIN BIKING THE SILK ROAD
Mountain bike from China to Pakistan with REI Adventures

Riding 70 to 100 kilometers a day, you will travel the Silk Road, bridging the Sino-CIS border. The route crosses from Kyrgyzstan to Kashgar in western China to the Karakorum Highway into Pakistan, along the rough and largely unpaved mountain roads of Kirgizia and Xinjiang. These regions have long been geographically and politically isolated from the rest of the world. After sightseeing in Samarkand, the first day of the ride will begin with a climb of the 12,000-foot Kegeti Pass, leading into the Karakol Valley. In the valley there are numerous herding settlements of yurts, the traditional wood-lattice and felt tents used by nomadic people across Central Asia and Mongolia. The rest of the 18-day ride will take you on a route along the peaks of the Tien Shan, the Kyokyo Meren Gorge, the Jumgal River and the Song-Kol Lake. You will continue your ride to the 11,580-foot-high Chatyr-Kol Lake, the Black Lake, the Subash Plateau, the Mintaka Valley, the Khunjerab Pass and the Batura Glacier. Rest days are distributed throughout the trip, giving you the option to helicopter over two of the highest peaks in the Commonwealth of Independent States and the Inilchick Glacier. There will also be ample opportunity to hike, visit local markets, explore historical sites and appreciate the endemic flora and fauna before you reach Pakistan and fly home.

ABOUT THE TRIP

LENGTH OF TRIP: 25 days

DEPARTURE DATES: June, July, Aug.

TOTAL COST: REI member $3,100-$3,500; non-member $3,200-$3,600

TERMS: Check, money order, MC, VISA; $400 deposit; cancellation fees depend on amount of advance notice

DEPARTURE: Fly to Samarkand, where you will meet the group.

AGE/FITNESS REQUIREMENTS: Must be at least 16 years of age; excellent physical condition

SPECIAL SKILLS NEEDED: Mountain biking experience; conditioning to prepare for high altitudes

TYPE OF WEATHER: Hot and dry at lower elevations, up to 90°; cold and windy at higher elevations; sun protection recommended

OPTIMAL CLOTHING: Detailed list is provided by REI.

GEAR PROVIDED: Communal items like tents and cooking equipment

GEAR REQUIRED: Mountain bikes, helmets, repair equipment, eating utensils, sleeping bag and pad

CONCESSIONS: Can purchase personal items and souvenirs before the trip begins in Tashkent

YOU WILL ENCOUNTER: Indigenous people (Kirghiz), unique wildlife, incredible mountain ranges, glaciers, remote wilderness

SPECIAL OPPORTUNITIES: Hiking, helicopter sightseeing, glaciers, photography

ECO-INTERPRETATION: Participants will learn from locals about the area they are visiting.

RESTRICTIONS: Women are discouraged from wearing shorts in cities.

REPRESENTATIVE MENU: *Breakfast*—porridge; *Lunch*—sandwiches; *Dinner*—stew; *Alcohol*—vodka and cognac available

ABOUT THE COMPANY

REI ADVENTURES
P.O. Box 1938, Sumner, WA 98390-0800, (800) 622-2236
REI Adventures, the travel division of REI, specializes in environmentally and culturally responsible travel around the world. The focus of its trips is active participation in small groups, with a philosophy that includes packing out all garbage and the use of a gas stove. The parent company, REI, is a consumer cooperative and has donated more than $2 million to conservation groups since 1976.

China

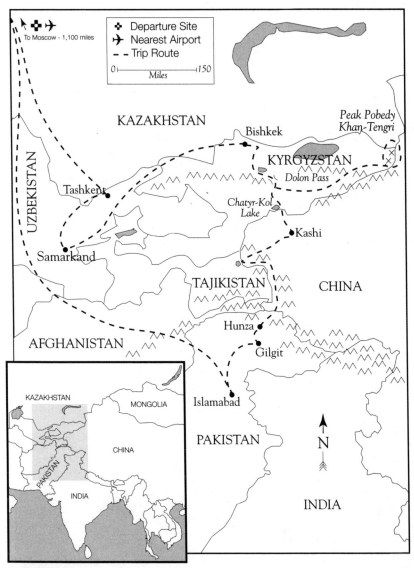

ECOFOCUS Overgrazing in western China was encouraged by the Communist government of the People's Republic of China to meet production quotas. These activities did not take environmental impact into consideration until the early 1990s. Problems include loss of arable land and erosion.

SOUTH CHINA BICYCLE TOUR
Bicycle through the countryside of southern China with REI Adventures

Traveling to a foreign country shouldn't be a whirlwind of airports and museums. You have come to experience life through the eyes of another. What better opportunity to come face to face with native Chinese people than to bicycle through their countryside? REI leads you on a journey through the magnificent landscape between Guangdong Province and Guangxi Autonomous Region, an agricultural area in the southern part of the country. Eleven distinct minority groups reside in the Guangxi Autonomous Region including the Yao and Miao. The climate is subtropical, and the landscape is dotted with conical limestone "karst" hills. The trip ends in Guilin, where karst hills tower above the lovely Xi River, resembling a Chinese landscape painting. Traveling by bike allows you to go at your own pace and stop for a cup of tea, if the mood strikes, or take the time to chat with people in farms, small villages or rural markets. All meals and accommodations will be in comfortable hotels, and the cycling will primarily be on flat to hilly dirt roads covered with gravel. A bilingual native Chinese guide will accompany you, and a "sag-wagon" can pick you up when necessary. Take the time to pedal through your vacation and really experience the fascinating culture and stunning scenery you traveled so far to see.

ABOUT THE TRIP

LENGTH OF TRIP: 17 days

DEPARTURE DATES: Nov. 11, 26; 1993: Feb. 3, 18

TOTAL COST: REI members: 11-17 people $1,195, 6-10 people $2,195. Nonmembers: 11-17 people $2,100, 6-10 people $2,310.

TERMS: VISA, MC, check, money order; $400 deposit; $25 cancellation fee if 180 days prior to departure

DEPARTURE: Arrive at airport in Hong Kong; transport to hotel provided

AGE/FITNESS REQUIREMENTS: Good physical condition and an active bicyclist

SPECIAL SKILLS NEEDED: Spirit of adventure and a positive attitude

TYPE OF WEATHER: Pleasant weather with cool temperatures, 45°-80°

OPTIMAL CLOTHING: Biking apparel

GEAR PROVIDED: 18-speed mountain bikes

GEAR REQUIRED: Personal gear

CONCESSIONS: Souvenirs available along route; bring personal items and film from home

YOU WILL ENCOUNTER: Indigenous peoples; some wildlife

SPECIAL OPPORTUNITIES: Learning and speaking Chinese

ECO-INTERPRETATION: Basic information about areas visited will be provided, including extensive cultural background of native people.

RESTRICTIONS: Riding permitted only on designated routes, but there will be opportunities to visit villages off the main route; no off-trail riding

MEALS: South China cuisine; ample water and beverages provided

ABOUT THE COMPANY

REI ADVENTURES
P.O. Box 1938, Sumner, WA 98390-0800, (800) 622-2236, (206) 395-5959
REI Adventures, the travel division of REI, specializes in environmentally and culturally responsible travel around the world. The focus of its trips is active participation in small groups, with a philosophy that includes packing out all garbage and the use of gas stoves. The parent company, REI, is a consumer cooperative and has donated more than $2 million to conservation groups since 1976.

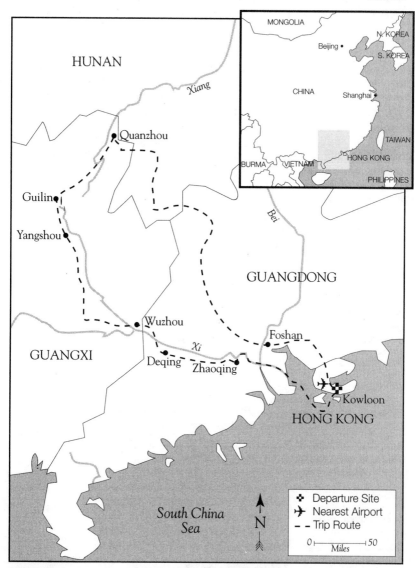

China

ECOFOCUS Industrial pollution of rivers and the taxing effects of a growing population are the People's Republic of China's largest environmental challenges. The adverse impact of feeding and housing an expanding population is being alleviated, but the Chinese presently cannot control the impact of industrial waste on their water, land and air.

CLASSIC TREK IN WEST PAMIR
Trek in the Pamir Mountains with Intertrek

Upon your arrival in Moscow you will visit Red Square, the Mausoleum of Lenin, the Cathedral of Basilius and the Kremlin. An early morning flight will take you to Dushanbe, the capital of Tajikistan. After visiting the bazaar and a traditional mosque you will enjoy a sumptuous Oriental dinner. The next few days will be spent traveling by buses and trucks into the spectacular Pamir Mountain region. The native Tajik crew will join you on the trek to hike five to six hours per day, looking at mountain terrain 16,000 to 23,000 feet high. You will be trekking through the Peter I (Krebet Petra Pervoga) mountain ridge. Following a route of steep ascents and descents you will find many sheep pastures and only two villages, populated mostly by Kirgis. The trek begins at 7,260 feet on a small plateau above the Muksu River. About 70 Kirgi families live in the first village you will reach. Crossing over glaciers, climbing unnamed passes and trekking next to crystal lakes, the route continues from one surreal landscape to the next. You will cross the Waisarek River on horseback and climb 10,560 feet to reach a steep descent over another river. After the trek, you will fly from Dushanbe to Samarkand, where you will have ample time to visit the bazaars and mosques before your departure.

ABOUT THE TRIP

LENGTH OF TRIP: 18 days
DEPARTURE DATES: May, June, July and Sept.
TOTAL COST: $2,370
TERMS: Check; $400 reservation, full balance due 60 days prior to trip; $50 fee for cancellations more than 180 days before trip, $200 fee 90-180 days before trip, $300 fee 60-89 days before trip
DEPARTURE: Local guides will meet the group in Moscow airport.
AGE/FITNESS REQUIREMENTS: Physically fit individuals capable of moderate activity
SPECIAL SKILLS NEEDED: None
TYPE OF WEATHER: Warm days (70°-95°) and chilly evenings (32° in mountains)
OPTIMAL CLOTHING: Hiking boots, lightweight pants, T-shirts
GEAR PROVIDED: Tents and equipment provided

GEAR REQUIRED: Sleeping bag
CONCESSIONS: Film and personal items unavailable
WHAT NOT TO BRING: Weapons, anything you can't carry
YOU WILL ENCOUNTER: Indigenous people (Tajiks, Ukrainians, Kirgis, Tartars, Mongolians and Russians)
SPECIAL OPPORTUNITIES: Theater and local festivals
ECO-INTERPRETATION: Guides will discuss environmental and cultural preservation issues as well as natural history.
RESTRICTIONS: No shorts in cities; do not photograph Tajiks without their permission or refer to them as Russians
MEALS: Foods are 90% local specialties

ABOUT THE COMPANY

INTERTREK
P.O. Box 50488, Phoenix, AZ 85046, (800) 346-4567
Specializing in adventure travel tours to central Asia, Intertrek, owned by Martin Hug, is a for-profit company that has been in business for 22 years with a staff of eight people. They use both local and American guides and stay in locally owned lodgings when they are not using tents. Intertrek works with the agricultural ministry in the CIS to facilitate growth in an environmentally sensitive manner by providing money through tourism.

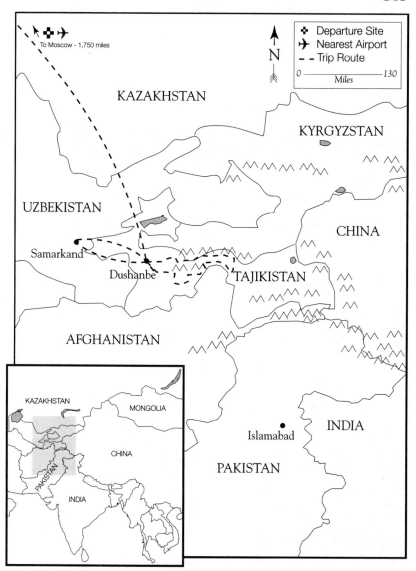

ECOFOCUS The proposed building of dams is a constant threat to the West Pamir region. An impending threat is a plan to flood a valley in this region for reasons that have yet to be clarified. The plans for the flooding were formulated 10 to 12 years ago, and only diligent efforts to preserve the region will stop the destruction of a unique wilderness area.

KAMCHATKA EXPLORATION
Backpack and raft in Siberia's Valley of Geysers with REI Adventures

In September 1990, the first Americans were granted permission to visit Kamchatka and the Valley of Geysers, the last undeveloped area of geyser concentrations in the world. Kamchatka, a land of 200 volcanoes (29 of them active), has long been shrouded in secrecy because of Soviet military bases in the area. It is protected as part of the Kronotsky Nature Preserve, part of the United Nations' network of nature preserves. The preserve was founded to protect the sable and its habitat. Only 100 permits are issued for foreign visitors each year, and the revenues from these permits go toward maintaining the park. Here you will see a unique geothermal phenomenon consisting of multiple tiers of geysers. Multicolored algae, boiling cauldrons of mud, azure blue pools and steam vents dot the landscape. For the next three days, you will backpack across the highlands of Kamchatka through dense patches of blueberries and wildflowers. At the Zhupanova River's convergence with the Lugovaya River, the rafting will begin! Endangered species such as Pacific and whitetailed eagles are easily seen along the river, along with ducks, falcons, grouse, caribou and Kamchatka brown bears. After tackling the rapids, you can spend your last day relaxing in a traditional Russian banya with a banquet of local specialties.

ABOUT THE TRIP
LENGTH OF TRIP: 12 days
DEPARTURE DATES: July 23, Aug. 27
TOTAL COST: $2,850
TERMS: VISA, MC, check, money order; $400 deposit; $25 cancellation fee 180 days prior to departure
DEPARTURE: Arrive Magadan, fly to Petropavlovsk and stay at tour base in Paratunka. Helicopter to wilderness.
AGE/FITNESS REQUIREMENTS: Active individuals in good health
SPECIAL SKILLS NEEDED: Experience hiking
TYPE OF WEATHER: Warm days with cool to cold evenings and mornings; temperatures 45°-65°
OPTIMAL CLOTHING: Rain clothes, layering dress
GEAR PROVIDED: Tents, rafts, dry-bags and community equipment, life jackets

GEAR REQUIRED: Insect repellent, raingear, packs, sleeping bag
CONCESSIONS: Bring all personal items and film from the US.
WHAT NOT TO BRING: Excess baggage
YOU WILL ENCOUNTER: Unique flora, fauna and geology of a restricted area
SPECIAL OPPORTUNITIES: Some of the best salmon fishing in the world and natural hot springs
ECO-INTERPRETATION: Effects of mining, geothermal power and tourist recreation are discussed.
RESTRICTIONS: None
REPRESENTATIVE MENU: *Breakfast*—porridge, Russian-style crepes, salmon; *Lunch*—smoked salmon with bread, salmon roe, crab, fried salmon, borscht ; *Dinner*—fish stew and lots of salmon dishes

ABOUT THE COMPANY
REI ADVENTURES
P.O. Box 1938, Sumner, WA 98390-0800, (800) 622-2236, (206) 891-2631
REI Adventures, the travel division of REI, specializes in environmentally and culturally sensitive travel around the world. REI Adventures trips operate in small groups and use kerosene stoves instead of wood, and all garbage is packed out. Revenue from the Kamchatka Expedition is used to help build and maintain boardwalks in the Valley of Geysers to protect fragile terrain. REI is a consumer cooperative that has donated more than $2 million to conservation groups since 1976.

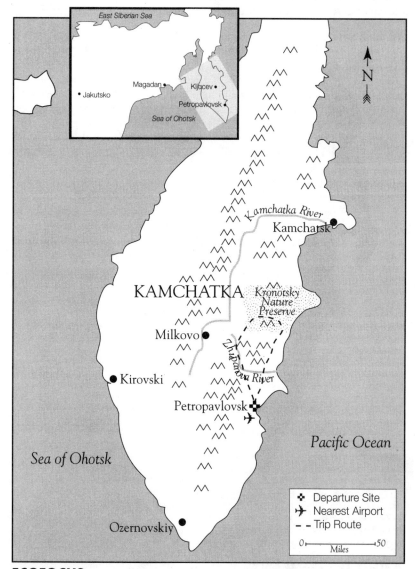

ECOFOCUS Kamchatka is a pristine wilderness area, untouched because the peninsula was declared off-limits to all foreigners and most Russians until 1990. Increasing pressures for recreational use and, more critically, development of geothermal power and mining threaten this unique wilderness area.

KURILE ISLANDS SAIL AND KAYAK
Explore the Kurile Islands with REI Adventures

The Kurile Islands are a natural paradise that have been jostled by the arms of a political battle between Russia and Japan for 45 years. Linking the Kamchatka Peninsula with Japan, the Kurile Islands form one section of the Pacific Ring of Fire. Consisting of active volcanoes, craters, hot springs, steep cliffs, blue lagoons and quiet bays, the islands are a sparsely settled, densely forested home to several rare species. The exploration begins with a tour of Yuzhno-Sakhalinsk, including a visit to a natural history museum, before you take a 20-hour ferry ride to Iturup Island. Camping, sailing and kayaking, you will travel across the Friza Straits and into small fishing villages. The route will allow you to explore uninhabited Urup Island and Tyatya, a 5,966-foot active volcano. On Yuzhno-Kurilsk, hot springs offer you a relaxing treat before you continue sailing to Shikotan Island, where you will camp on Point "Kraisveta," translated as "the point at the edge of the world." The schedule is flexible and will change as discoveries are made about this relatively unexplored area. With an open mind and a zest for travel, you will enjoy this incredible opportunity to explore an until-recently restricted area and to interact with a unique blend of cultures.

ABOUT THE TRIP
LENGTH OF TRIP: 16 days
DEPARTURE DATES: Sept. 1
TOTAL COST: REI members: $2,450, 10-12 people; $2,750, 5-9 people
TERMS: VISA, MC, check, money order; $400 deposit; $25 cancellation fee if notification is given 180 days in advance
DEPARTURE: Arrive in Yuzhno-Sakhalinsk on the Island of Sakhalin.
AGE/FITNESS REQUIREMENTS: Active individuals in good health
SPECIAL SKILLS NEEDED: A spirit of adventure; sea kayaking experience is an asset but not required
TYPE OF WEATHER: Chance of fog and rain; Aug.-Sept. is the sunniest time of year with temperatures 40°-70°
OPTIMAL CLOTHING: Clothing list is provided
GEAR PROVIDED: Tents, kayaks, dry-bags and community camping equipment are provided.

GEAR REQUIRED: Sleeping bags and personal gear
CONCESSIONS: Purchase all personal items and film in advance.
YOU WILL ENCOUNTER: Contrasting cultures of Russian, Korean and Japanese ancestry; wide variety of sea and shore life, including the far east skunk and kunahir grass snake
SPECIAL OPPORTUNITIES: Optional volcano hike, hot springs and fishing opportunities
ECO-INTERPRETATION: Discussion of the political and natural history of the Kurile Islands and the endemic wildlife will take place.
RESTRICTIONS: Will be discussed by guides
REPRESENTATIVE MENU: *Breakfast*—porridge, crepes; *Lunch*—fish, borscht, breads; *Dinner*—fish stew; *Alcohol*—no policy

ABOUT THE COMPANY
REI ADVENTURES
P.O. Box 1938, Sumner, WA 98390-0800, (206) 891-2631, (800) 622-2236
REI Adventures, the travel division of REI, specializes in environmentally and culturally responsible travel around the world. The focus of its trips is active participation in small groups, with a philosophy that includes packing out all garbage and the use of gas stoves. The parent company, REI, is a consumer cooperative and has donated more than $2 million to conservation groups since 1976.

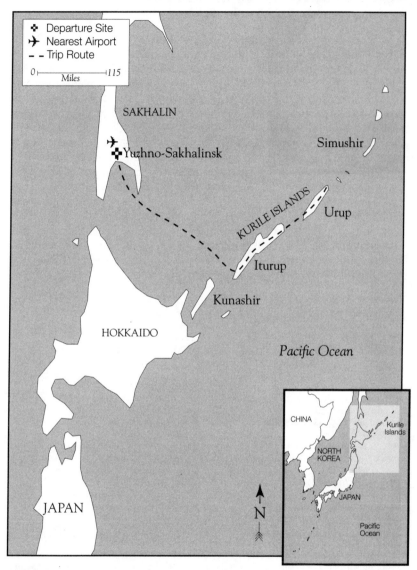

ECOFOCUS The Kurile Islands have experienced very limited outside visitation and are sparsely inhabited. The indigenous cultures and fragile environment are susceptible to the pressures of increased visitation as these areas are opened to tourism. In a country faced with the challenges of change and the need for hard currency, responsible tourism will support the area's natural resources.

SEA KAYAKING IN LAKE BAIKAL
View wildlife in Siberia with Baikal Reflection, Inc.

Surrounded by five mountain ranges and fed by 336 tributaries, Lake Baikal is an incredible geological dimension containing diverse and largely endemic life forms. The unique flora and fauna of this area includes the world's only freshwater seal. After two days of travel by plane and boat, you will arrive in Komsomalsk na Amure, where you will board the Baikal Amur Mainline for two nights. The trip begins after a night in local homes near Verikh Zayimka. You will kayak into areas like the Hakusi Hot Springs, the Selenga Delta, the Barguzin River, Yarki Island and the Upper Angara Delta. Each route is used only one time each year to minimize the group's impact. Regardless of the path you follow, you are guaranteed to be immersed in the splendor of this scenic area. Travel along the shores of Lake Baikal on rivers that will lead you through taiga forests of birch, larch, cedar and pine. You will have the opportunity to explore the river deltas and estuaries for the interesting and diverse wildlife of the area. The return route will take you to Ulan Ude to visit the Buryat Ethnological Museum, Museum of Natural History and the Buddhist temple, Ivolginskaya Dashan, before you stay with local Muscovites and independently tour the city.

ABOUT THE TRIP

LENGTH OF TRIP: 15 or 17 days

DEPARTURE DATES: June, July and Aug.

TOTAL COST: $1,500-$1,700

TERMS: VISA, MC and cash; deposit of $500; cancellation fee of $75 to 100% of land cost relative to notification date

DEPARTURE: Arrive in Khabarovsk accompanied by an interpreter and be taken to your stay at a Siberian home or hotel.

AGE/FITNESS REQUIREMENTS: Medical clearance required

SPECIAL SKILLS NEEDED: None

TYPE OF WEATHER: Warm, humid climate with chance of rain; 40-80°

OPTIMAL CLOTHING: Light activewear, cool evening wear, raingear

GEAR PROVIDED: Ocean kayaks

GEAR REQUIRED: Gear list provided

CONCESSIONS: Not usually available; bring supplies with you

WHAT NOT TO BRING: Items that display wealth

YOU WILL ENCOUNTER: Indigenous people (Evenk, Yakut, Buryat), freshwater seals, Siberian tigers, bird life

SPECIAL OPPORTUNITIES: Fishing, hot springs, assistance in environmental work

ECO-INTERPRETATION: Preservation issues will be discussed.

RESTRICTIONS: Respectful conduct expected

REPRESENTATIVE MENU: *Breakfast*—bread, cucumbers and tomatoes; *Lunch*—cucumbers, tomatoes, potatoes and bread; *Dinner*—fish, sausages and vegetarian options; *Alcohol*—available, but not encouraged

ABOUT THE COMPANY

BAIKAL REFLECTION, INC.
13 Ridge Rd., Fairfax, CA 94930, (415) 455-0155, (800) 927-2797
Working with different cultural ideas and economic systems to benefit local environments, Baikal Reflection runs trips to Siberia. The company contributes to studies on the local freshwater seal and uses each travel route once per year. Baikal Reflection, owned by Megan Phillips-Boyle and Fred Boyle, is a for-profit company with a staff of ten that has been in business for two years.

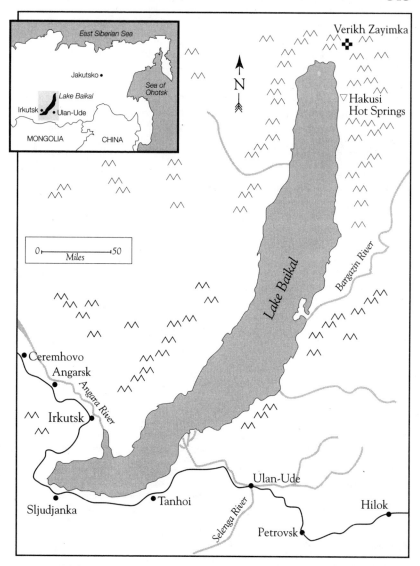

ECOFOCUS The delicate ecosystems of the Lake Baikal area are threatened by the Baikal Amur Mainline (BAM) Railway Commission's plans to encourage mining, timber and other extractive industries to support the railroad. As a result, natural resources such as the taiga subarctic forest are in danger of being exploited. Lake Baikal is threatened by pollution upstream. The Siberian tiger is one of many species being hunted illegally.

HIMALAYAN RUN AND TREK
Travel the Himalayas with Force 10 Expeditions Ltd.

On the Himalayan Run and Trek, you will have the opportunity to run or walk at your own pace along with local athletes. Your participation will help focus attention on the pressing need to conserve the Himalayan environment, provide funding for local efforts already under way. After arrival in Delhi, the first event is a five-kilometer race right up to the steps of the Taj Mahal in Agra. Proceeds will be donated to victims of the October 1991 earthquake in India. Flying to Darjeeling, you begin a five-day walking tour, which is highlighted by an international running event suitable for both elite and casual runners. With views of Mt. Everest, Kanchenjunga and mountains in five countries, the run and trek is designed to inspire all participants to a higher awareness of the Himalayan environment. Later, in Sikkim, you visit ancient monasteries and can join hundreds of local runners in the Gangtok Half Marathon, another charity race. This trip is perfect for marathon and casual runners or non runners interested in visiting places off the beaten track and anxious to help preserve one of the most extraordinary and beautiful areas of the world. Many more treks are available in this area either before or after the Himalayan Run and Trek.

ABOUT THE TRIP

LENGTH OF TRIP: 16 days

DEPARTURE DATES: Oct. 24

TOTAL COST: $2,200

TERMS: Check; $300 nonrefundable deposit withheld upon cancellation.

DEPARTURE: Arrive at airport in Delhi; pre-arranged taxi to hotel

AGE/FITNESS REQUIREMENTS: Healthy enough to walk

SPECIAL SKILLS NEEDED: Ability to run or walk a half-marathon

TYPE OF WEATHER: Little or no rain; 70° in the daytime, 40° at night

OPTIMAL CLOTHING: Running shoes, lightweight hiking boots, shorts

GEAR PROVIDED: Aid stations for runners, jeep support, guides and porters, cooks, all food except four meals in Delhi

GEAR REQUIRED: Duffel, sleeping bag

CONCESSIONS: Film, personal items and souvenirs available in towns along the way

WHAT NOT TO BRING: Camping equipment, fancy clothes

YOU WILL ENCOUNTER: Many indigenous peoples during 5-day stage run and trek, local athletes, Sherpas; wildlife such as the red panda and one-horned rhino

SPECIAL OPPORTUNITIES: Running

ECO-INTERPRETATION: History of Himalayan exploration is provided at the only Himalayan museum in the world.

RESTRICTIONS: Permit to visit Sikkim required; respect in monasteries

REPRESENTATIVE MENU: *Breakfast*—eggs, cereal, fruit; *Lunch*—sandwiches, potatoes, fruit; *Dinner*—meat, pasta, fruit; *Alcohol*—smoking and alcohol discouraged

ABOUT THE COMPANY

FORCE 10 EXPEDITIONS LTD.
P.O. Box 34354, Pensacola, FL 32507, (800) 922-1491

In business 20 years, Force 10 Expeditions Ltd. is a for-profit organization owned and operated by Jim Crosswhite. The company uses local guides and hotels and works to educate Sikkim people about preservation through cross-cultural exchanges. Also, a portion of the proceeds goes toward running shoes for local athletes with scholastic and athletic ability, and climbing equipment is given to Indian climbers through the Himalayan Mountaineering Institute. Donations are also raised for victims of the earthquake in India and other causes.

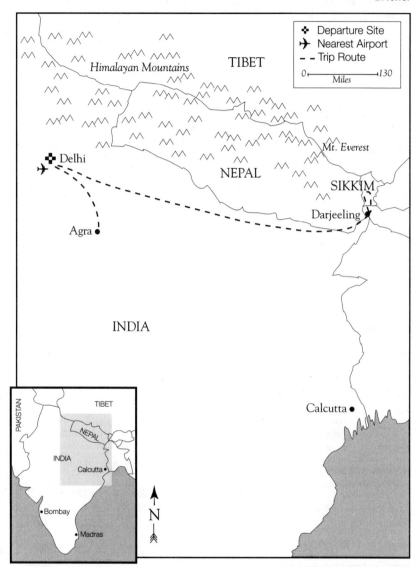

ECOFOCUS The eastern Himalayas are largely closed to mountaineers and tourists, but small areas are opening slowly. Darjeeling, Gangtok and Western Sikkim are culturally interesting, with important monasteries and trekking routes. Government officials are open to conservation ideas and welcome support.

ZANSKAR AND LADAKH

Explore Buddhist monastery culture in India with Himalayan High Treks

This extraordinary trek takes you high in the Indian Himalayas on the Tibetan Plateau, an area much like ancient Tibet. Here you have the opportunity to experience the religious rituals and the monastery-centered lifestyles that have disappeared from Chinese-occupied Tibet. For 22 days, you will hike through spectacular mountain scenery, crossing several passes (four of which are over 16,000 feet in elevation) and numerous rivers. Ponies carry your gear, and the camp staff puts up the tents and prepares the meals, leaving you free to enjoy the fascinating culture and magnificent views. You will visit the colorful Zanskaris, the ethnic Tibetans who inhabit the Zanskar Valley, and Lingshot Gompa, home to 60 monks and known for its school of painting. You will see the oldest monastery in Ladakh, Lamayuru, built in the 10th century, which today boasts a large library of Buddhist literature and is famous for its festivals. The trip begins in Delhi and includes a visit to the Taj Mahal and the Hindu shepherding communities of the fertile Kulu Valley. The trek is scheduled early in the season, an excellent time for wildflower viewing. Prepare to exert yourself and be richly rewarded for your efforts with unsurpassed scenery and valuable cultural history.

ABOUT THE TRIP

LENGTH OF TRIP: 32 days

DEPARTURE DATES: July 1

TOTAL COST: $2,250

TERMS: Cash, check; $250 deposit; loss of deposit if cancellation occurs more than 60 days prior

DEPARTURE: Meet guide upon arrival at airport in New Delhi

AGE/FITNESS REQUIREMENTS: All ages; good health and fitness

SPECIAL SKILLS NEEDED: Experienced hikers

TYPE OF WEATHER: Clear, cool weather during the day in the 60s and 70s; comfortably cool at night in the 40s and 50s.

OPTIMAL CLOTHING: Layered clothing, lightweight hiking boots

GEAR PROVIDED: Tents, kitchen and cook, first-aid kit and nurse, group gear

GEAR REQUIRED: Sleeping bag and ground pad

CONCESSIONS: Bring film from home; souvenirs and personal items can be purchased along the way.

YOU WILL ENCOUNTER: Constant contact with indigenous people; wildlife includes Himalayan black bears, marmots and many types of birds

SPECIAL OPPORTUNITIES: Meditation and pujas are open to visitors at many monasteries.

ECO-INTERPRETATION: Information given about groups like the Ladakh Foundation, which promotes responsible development in the area and where and how to advocate conservation

RESTRICTIONS: Must be properly dressed to enter villages and monasteries; briefing will be provided

REPRESENTATIVE MENU: *Breakfast*—tea, coffee, tsampa porridge, chapatis, eggs, fruit; *Lunch*—sandwiches with cheese and chutney, hot chocolate or tea; *Dinner*—soup, rice with curry, vegetables; *Alcohol*—chang (a local rice beer) available after acclimation to altitude

ABOUT THE COMPANY

HIMALAYAN HIGH TREKS
241 Dolores St., San Francisco, CA 94103, (415) 861-2391

For three years, Himalayan High Treks has been a for-profit company specializing in Himalayan trekking with a strong cultural emphasis. Effie Fletcher, the director, has been a Sierra Club leader for years and leads the trips herself along with eight other guides. They have helped with local projects, including building bridges and village cleanups. Guests on their trips are taken to visit a solar center in Leh, Ladakh. They also cooperate with Bay Area Friends of Tibet, supporting the Dalai Lama's peace efforts by using related businesses and making donations.

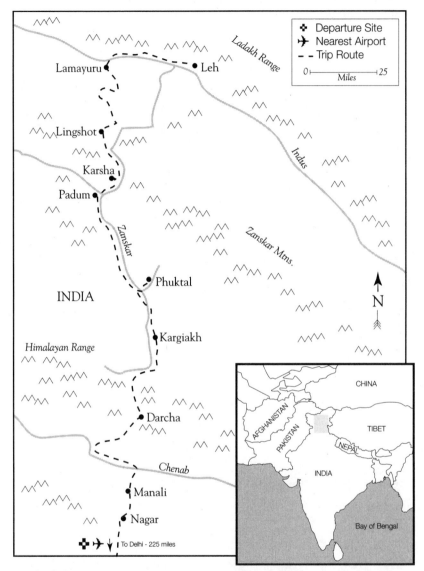

ECOFOCUS Water quality is a great environmental concern in this area, due to the Chinese dumping nuclear waste in Tibet. Poor sanitation also degrades water quality. In addition, forests are suffering from air pollution caused by automobiles and factories.

MOUNTAINS, TEMPLES AND HAMLETS
Trek through the mountains and villages of Japan with Journeys East

Beginning in Tokyo, where modern technology mixes with centuries-old narrow streets, Journeys East will take you from the city through the countryside and into the mountains. You will spend a day on the slopes of Mt. Fuji, then continue trekking into the spectacular Kita Alps. For five days, you will trek from lodge to lodge through this mountainous area in central Japan. The route will pass through alpine forests and craggy mountain bowls. The villages of thatched roofs, open hearths and hand-hewn beams are home to skilled artisans, making traditional pottery and lattice-work carvings. The simplicity of the farming villages and the intricate art-work are inspiring. Generations of Takayama villagers are famous for their woodwork, inspired by the Kita Alps and represented by the shrines the Imperial Court orders to be built around the country. The wildlife along the trek is exquisite and includes mountain serows, monkeys and ptarmigans. Your journey will come to an end in Kyoto, where you will stay in an exquisite ryokan (inn) with tranquil gardens in the heart of the inner city. The last few days are flexible, allowing you to explore the cornucopia of villas, gardens, temples and palaces that beautifully illustrates Japanese culture.

ABOUT THE TRIP

LENGTH OF TRIP: 16 days

DEPARTURE DATES: May, Sept.

TOTAL COST: $3,125

TERMS: Checks; $200 deposit; cancellation policy varies according to time of notification

DEPARTURE: Arrive in Tokyo's Narita airport and guides will escort group by express train to Tokyo.

AGE/FITNESS REQUIREMENTS: Open minds and flexible legs

SPECIAL SKILLS NEEDED: Some hiking experience

TYPE OF WEATHER: Temperatures ranging from mild days to cold nights, 30°-70° in the spring; the crisp fall will include temperatures of 30°-60°

OPTIMAL CLOTHING: A list will be provided by the outfitter.

GEAR PROVIDED: Handouts outlining cultural history

GEAR REQUIRED: A list will be provided by the outfitter.

CONCESSIONS: Film, personal items and souvenirs can be purchased in Tokyo.

WHAT NOT TO BRING: Too much gear

YOU WILL ENCOUNTER: Indigenous people, birds, monkeys, serows, tropical fish

SPECIAL OPPORTUNITIES: Farmhouse living, Zen meditation, indigo dyeing and pottery

ECO-INTERPRETATION: Discussion of the effects of technology and pesticides in Japan

RESTRICTIONS: Travelers must conform to Japanese traditions.

REPRESENTATIVE MENU: *Breakfast*—rice, soup, salad, fish; *Lunch*—noodles, sushi, tempura; *Dinner*—Japanese cuisine; *Alcohol*—available at all meals

ABOUT THE COMPANY

JOURNEYS EAST
2443 Fillmore St. #289, San Francisco, CA 94115, (510) 601-1677
A for-profit company owned by Davis Everett, Journeys East has been in business for eight years with a staff of three. The company specializes in culturally oriented trips to Japan during which local lodgings are used and trail maintenance is performed. Journeys East uses trained anthropologists and naturalists as guides.

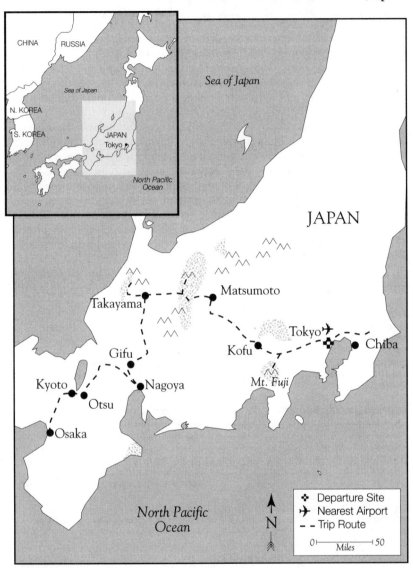

ECOFOCUS Animals like the whale and the dolphin are threatened due to the fishing industry in Japan. The coast is lined by industry that contributes to air, land and water pollution. Deforestation and waste disposal are also issues of concern in this country.

MALAYSIAN TROPICAL ADVENTURE TREK
Hike, boat and swim through rainforests with Asian Pacific Adventures

Malaysia's national parks are home to vast amounts of hidden treasures including rafflesia, the biggest flower in the world, 300 species of birds and more than 1,000 varieties of orchids. Among the magnificent gorges, racing rivers and tremendous hills awaits the adventure of a lifetime. This expedition will take you whitewater rafting, wildlife tracking, birdwatching, hiking, swimming in calm river waters, exploring limestone caves and camping amidst towering tropical trees. Your trek will begin on the island of Borneo in Sabah, a Malaysian province, with an ascent of 13,455-foot Mt. Kinabalu, Southeast Asia's highest mountain. You will visit a wildlife sanctuary where orphaned baby orangutans are reintroduced to the wild. Observe some of the orphans at feeding time in the jungle and see them being trained to climb vines. While in Sabah, meet indigenous people in their communal longhouses and examine their exquisite folk crafts. Or just relax and snorkel off the shores of pristine Sapi Island. Then return to the heart of peninsular Malaysia and Taman Negara National Jungle Reserve and Park, a vast rainforest estimated to be 130 million years old. In this land of many surprises, this trek will reveal the flora and fauna that dreams are made of and will show you a good time along the way.

ABOUT THE TRIP
LENGTH OF TRIP: 16 days
DEPARTURE DATES: Jun. 25, Jul. 9, Aug. 6, 20
TOTAL COST: $2,176
TERMS: Cash, check; Minimum $300 deposit; $300 cancellation fee less than 90 days prior
DEPARTURE: Arrive in Kuala Lumpur, where you will meet up with the group.
AGE/FITNESS REQUIREMENTS: Minimum 16 years old without parent, good health, able to hike four hours a day
SPECIAL SKILLS NEEDED: None
TYPE OF WEATHER: Hot and humid; count on one cold morning at 13,455 ft. above sea level
OPTIMAL CLOTHING: Loose-fitting clothing: hand-washable T-shirts, shorts and jeans; sturdy footwear (hiking boots) and a double layer of socks
GEAR PROVIDED: Sleeping bags
GEAR REQUIRED: Rucksack/ backpack/ fanny pack
CONCESSIONS: Personal items, souvenirs and film are available on the trip.

WHAT NOT TO BRING: Too many clothes
YOU WILL ENCOUNTER: Kadazan and Rungu people on Borneo, Malays and aboriginal people in the jungle on peninsular Malaysia; excellent wildlife including hornbills, drongos, kingfishers, warblers, bulbuls, elephants, monkeys, tapirs, gibbons, otters, bats, etc.
SPECIAL OPPORTUNITIES: Hiking, easy rafting, swimming, snorkeling and scuba
ECO-INTERPRETATION: Discussion of indigenous peoples and wildlife provided by local guides
RESTRICTIONS: Many Malays are Muslim; travelers must show respect.
REPRESENTATIVE MENU: *Breakfast*—anything in cities; eggs, cereal, breads, fruit on trek; *Lunch*—anything in cities; sandwiches, fruit, hot noodles on trek; *Dinner*—anything in cities; rice and Malaysian food on trek; *Alcohol*—not available on trek, but can be purchased in cities

ABOUT THE COMPANY
ASIAN PACIFIC ADVENTURES
826 S. Sierra Bonita Ave., Los Angeles, CA 90036, (213) 935-3156, (800) 825-1680 (Outside California)
The mission of Asian Pacific Adventures is to provide culturally and ecologically sensitive, educational and fun trips that give participants an understanding of remote and fragile regions and cultures. In business seven years, this is a for-profit company that uses local guides and contributes money to local economies. It is a founding member of Specialty Travel Alliance and participates in trips that include low-impact camping.

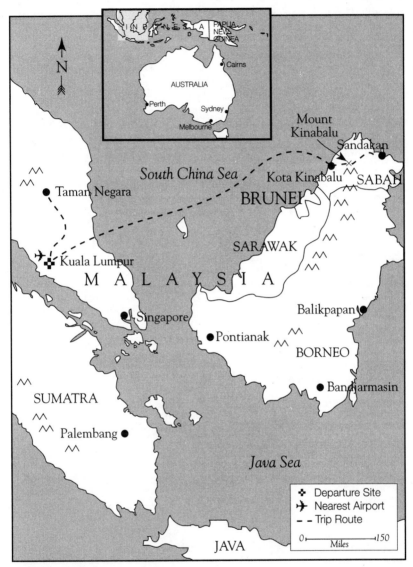

ECOFOCUS Malaysia is threatened by the impending destruction of a 130-million-year-old rainforest, which is home to an incredible variety of plants, wildlife, insects and indigenous people. Many of these are unique to this region and face extinction due to the destruction of the rainforest through extensive logging, dams and encroaching rubber plantations.

KARNALI RIVER TREK
Raft and trek in Nepal with Cascade Raft Co.

Located in the remote western section of Nepal, this premier stretch of whitewater rollicks through big Class IV and V rapids. The Karnali River, one of the longest in Nepal, is so secluded that eight days of trekking are required to reach the put-in. The excursion takes trip participants through a cross section of the small kingdom—25 distinct ethnic groups and their architecture, dress and customs–and a striking variety of plants, animals and birds that coexist in a land the size of North Dakota. You will cover all of the major geographical areas of Nepal, including the scenic Kathmandu Valley, the world's highest peaks in the Himalayas and the rarely rafted Karnali River. The trekking portion of the trip takes the group through high mountain villages where people still use primitive tools to build their houses, make their clothes with homegrown wool and harvest and grind grain by hand. The first three days of rafting will include many thunderous rapids, some of which will be run and some of which will be portaged. Along the way, the group will stop at villages to sample the indigenous food and view crocodiles, monkeys and exotic birds.

ABOUT THE TRIP

LENGTH OF TRIP: 28 days
DEPARTURE DATES: Nov. 23
TOTAL COST: $2,600
TERMS: Cash; $500 deposit to reserve; full refund if made 60 days or more prior to trip
DEPARTURE: Nearest airport in Kathmandu; shuttle provided to Kathmandu departure point
AGE/FITNESS REQUIREMENTS: Good physical condition
SPECIAL SKILLS NEEDED: Class IV whitewater rafting experience
TYPE OF WEATHER: Warm temperatures (70°-85°), cold water
OPTIMAL CLOTHING: Lightweight, layered clothing
GEAR PROVIDED: Tents and river gear
GEAR REQUIRED: Sleeping bag and pad, backpack, day pack
CONCESSIONS: Film and personal items should be purchased prior to departure. Souvenirs are available on trail and after trek.

WHAT NOT TO BRING: Formal attire
YOU WILL ENCOUNTER: Excellent whitewater, scenery and new culture
SPECIAL OPPORTUNITIES: Mountain biking, trip to Tiger Top with trip extension
ECO-INTERPRETATION: The guides are knowledgeable about the area and give talks about indigenous people.
RESTRICTIONS: Nepal is a conservative society; must show cultural sensitivity
REPRESENTATIVE MENU: *Breakfast*—eggs, local breakfast; *Lunch*—sandwiches, fresh fruit, tuna, cheese, vegetables, peanut butter, vegetable spreads; *Dinner*—chicken mixed with vegetables, lentils, pasta, fried potatoes, baked desserts, pudding and cheesecake; *Alcohol*—some alcohol allowed, but must be purchased in Kathmandu

ABOUT THE COMPANY

CASCADE RAFT CO.
P.O. Box 6, Garden Valley, ID 83622, (800) 292-7238
In operation for eight years, Cascade Raft Co. is a for-profit outfitter with a staff of 35, specializing in whitewater and trekking trips to Idaho, Nepal and Costa Rica. Owned by Steve Jones, the company is also active in protecting local and international rivers from dams and other damaging projects through Idaho Rivers United and Project Raft. Steve Jones and Nema Lama, a Nepalese river guide, are also dedicated to protecting rivers in Nepal from the threats imposed by dams like the Marysadi Dam, which was built despite the opposition demonstrated.

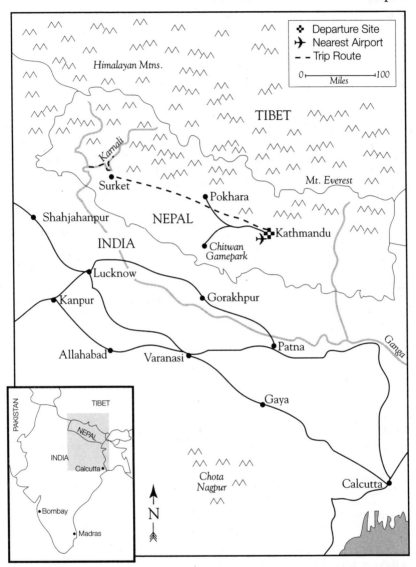

ECOFOCUS Overpopulation is a pressing issue in Nepal and is one of the underlying causes of environmental problems such as depletion of forests and erosion of arable land. The farm-based rural economy results in an abundance of rice paddies on steep slopes, which weakens the soil. Efforts are being made to rotate crops to stop erosion and landslides.

THAILAND: A TROPICAL ADVENTURE
Walk, boat, swim and explore Thailand with Bolder Adventures

This special trip was designed by adventure travel experts who have lived in Thailand for years and picked what they consider to be the very best experiences available to visitors of this magical country. On this exploration, you will experience the height of the physical beauty and cultural diversity of Thailand—from the pristine, uninhabited tropical islands to life among native people in remote tribal villages. In Bangkok, you will stay close to the Chao Phraya River and travel by river taxis, the most enjoyable and scenically rewarding way to experience this fascinating, bustling city. In ChiaPng-Mai, the group will stay in a complex of luxurious teak villages owned by a local family. Opportunities abound to cater this extraordinary vacation to your own personal desires. For those seeking more challenge in their trip, enjoy hiking, elephant riding and bamboo rafting in the rugged jungle-covered mountains. Noted lecturers are available to facilitate your understanding of this complex country. By participating in this voyage, you will explore the essence of Thailand while making a contribution to its preservation and environmental well-being.

ABOUT THE TRIP

LENGTH OF TRIP: 16 days, with optional extensions available

DEPARTURE DATES: Jul. 11, Nov. 7, Dec. 12 in 1992; Jan. 16, Feb. 6 in 1993

TOTAL COST: $2,295

TERMS: Cash, check; $300 VISA or MC deposit; cancellation fee depends on amount of advance notice

DEPARTURE: Arrive at airport in Bangkok; special connection fares provided

AGE/FITNESS REQUIREMENTS: Good health

SPECIAL SKILLS NEEDED: None

TYPE OF WEATHER: Warm and dry; 55°-85°

OPTIMAL CLOTHING: Cool, light cottons, sturdy walking shoes

GEAR PROVIDED: Sleeping bag, sleeping pad, mask, fins, snorkel

GEAR REQUIRED: None

CONCESSIONS: Film, personal items and souvenirs can be purchased in major cities and villages along the way.

WHAT NOT TO BRING: Excess baggage, electrical appliances

YOU WILL ENCOUNTER: Hill tribe indigenous peoples; limited wildlife viewing

SPECIAL OPPORTUNITIES: Snorkeling, hiking, scuba, rafting, swimming, sea kayaking

ECO-INTERPRETATION: The trip leader is an American with a background in biology and anthropology who has lived in Thailand 10 years.

RESTRICTIONS: Many; cultural primer is provided for participants

REPRESENTATIVE MENU: *Breakfast*—eggs, toast, fresh fruit, coffee, tea; *Lunch*—rice, vegetables, curry dishes, noodles; *Dinner*—chicken, seafood, rice, Thai curries, soup, noodles, fresh fruit; *Alcohol*—available for purchase

ABOUT THE COMPANY

BOLDER ADVENTURES
P.O. Box 1279, Boulder, CO 80306, (800) 397-5917, (303) 443-6789
Aiming at getting travelers close to the people, cultures and natural history of Southeast Asia, Bolder Adventures is a for-profit adventure travel organization in business five years. Its activities in Thailand are focused on cultural preservation of the tribal society of the north, and helping them become educated about how to lessen the impact of their traditional slash-and-burn agriculture. This company supports reforestation, local schools, water projects and latrine building in hill tribe areas by donating books and other supplies.

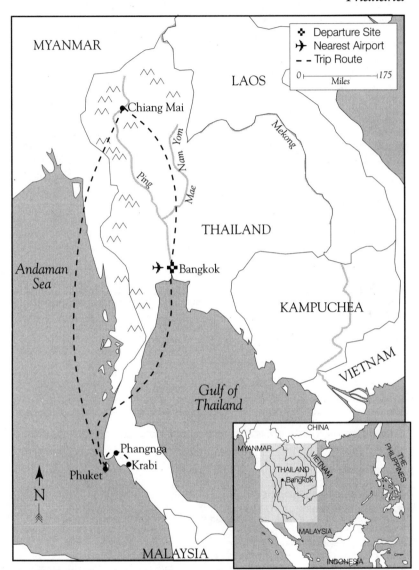

ECOFOCUS Thailand's major environmental threat is its population explosion; 55 million people live in a landmass the size of France. Thousands of tribal people experiencing persecution in China, Laos, Myanmar and Vietnam have migrated to Thailand. Subsistence living produces high environmental impact due to slash-and-burn agriculture, although there is currently a ban on all logging.

CAMEL SAFARI IN THE RED CENTER
Ride camels and hike in Australia with Outer Edge Expeditions

The Australian Red Center is an area of uncharted land that is inhabited only by wildlife. The Macdonnell Ranges are located here, as well as vermilion and cream-colored cliffs of sandstone, comprising one of the largest mountain systems in Australia. The camel safari centers around the foothills and dry riverbeds of the range, which is filled with gorges, valleys and water holes. Examine ancient Aboriginal rock paintings, view scenic gorges, explore unusual rock formations and riverbeds while experiencing a nomadic lifestyle. Swim in a water hole and gallop across the sand on your camel. Late each day, you will return to base camp and turn your camel out to graze. The evening meal is cooked on an open fire around which Aboriginal legends are exchanged. Sleep at night under a canopy of stars in a swag—a mattress and pillow in a canvas cover, which can double as a comfortable seat. Wake to the beauty of another day exploring the flora and fauna of the outback. On the last day you will return to Alice Springs, where you will have time to see the town before you catch a flight back to Sydney for your final departure.

ABOUT THE TRIP

LENGTH OF TRIP: 11 days

DEPARTURE DATES: March, Aug. and Oct.

TOTAL COST: $890

TERMS: Check; $300 deposit with no cancellation penalty up to 2 months before departure

DEPARTURE: Arrive in Sydney and transfer onto a flight to Alice Springs.

AGE/FITNESS REQUIREMENTS: Moderate fitness

SPECIAL SKILLS NEEDED: A sense of adventure

TYPE OF WEATHER: Sunny, hot and dry; up to 90° in the day, down to 60° at night

OPTIMAL CLOTHING: Light long-sleeve shirts, long pants, wide-brim hat, sneakers

GEAR PROVIDED: All required gear is provided.

GEAR REQUIRED: None

CONCESSIONS: Film, souvenirs and personal items can be purchased in Alice Springs.

WHAT NOT TO BRING: Excess baggage

YOU WILL ENCOUNTER: Desert wildlife: euro kangaroos, brumbies (wild horses), dingos, wild camels and colorful birds

SPECIAL OPPORTUNITIES: Fantastic photographic opportunities, camel riding, nomadic lifestyle

ECO-INTERPRETATION: Expedition leader will discuss local flora, fauna and culture

RESTRICTIONS: None

REPRESENTATIVE MENU: *Breakfast*—bush damper and eggs; *Lunch*—sandwiches; *Dinner*—casseroles; *Alcohol*—not encouraged

ABOUT THE COMPANY

OUTER EDGE EXPEDITIONS
45500 Pontiac Trail, Ste. C, Walled Lake, MI 48390, (800) 322-5235, (313) 624-5140
Specializing in small-group expeditions worldwide, Outer Edge Expeditions, owned by Brian Obrecht, is a for-profit business with a staff of 11. The outfitter, in operation for four years, uses local resources and focuses on outdoor education. Outer Edge Expeditions also donates a portion of its proceeds to the Nature Conservancy for conservation activities in central Australia.

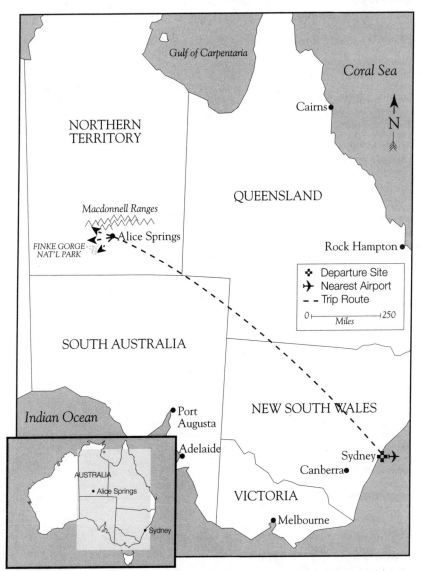

ECOFOCUS Introduction of Western culture to the indigenous Aboriginal culture has had serious side effects. The younger Aboriginals are forsaking their families, culture and lifestyle for Western material goods. The culture as a whole is being threatened, as is Aboriginal knowledge of the environment. The desert itself is changing, since Aboriginals are no longer burning it at regular intervals as they have for 40,000 years.

WILDLIFE OF THE GREAT BARRIER REEF
Dive, snorkel and hike in Queensland with Biological Journeys

Explore Australia while building a solid understanding of the complexities and importance of conservation. Snorkel along the Great Barrier Reef, a 1,250-mile-long barrier of underwater coral gardens filled with marine life. Awaken to the shrill cry of a sulphur-crested cockatoo, and fly over the Whitsunday Islands, looking down upon a maze of coral formations. Live on a boat among the complex outer Barrier Reef formations and around the numerous islands closer to the shore. The water here is filled with clownfish darting in and out of anemone and unicorn fish exploring sea fans. You will investigate reefs with names like Hook, Line, Bait and Sinker. If the clouds hide the afternoon sun, you may be lucky enough to see the nocturnal coral polyps feeding. Upland, the rainforests of Eungalla National Park, Kangaroo Island and the dry eucalyptus outback of northern Queensland are awaiting you. Unique marsupials of Australia, as well as a spectacular representation of the varied and beautiful bird life, can be seen throughout the trip. You will visit the Atherton Tablelands, which are among the richest agricultural areas of Australia. When in Cairns, you will travel to the Daintree River and the Kuranda Butterfly Sanctuary.

ABOUT THE TRIP

LENGTH OF TRIP: 22 days

DEPARTURE DATES: Oct., Nov.

TOTAL COST: $5,195

TERMS: Check; deposit of $500; cancellation fee of $100 with 60 days' advance notice

DEPARTURE: Arrive in Cairns.

AGE/FITNESS REQUIREMENTS: Average physical fitness

SPECIAL SKILLS NEEDED: A love of nature

TYPE OF WEATHER: Warm and dry 75°-85° daytime temperatures with cooler evenings

OPTIMAL CLOTHING: Casual camping clothes and tennis shoes

GEAR PROVIDED: Bedding, towels and diving equipment

GEAR REQUIRED: Collapsible duffel bag, snorkeling gear

CONCESSIONS: Available when not in outback or on boat

WHAT NOT TO BRING: Hard suitcases or formal attire

YOU WILL ENCOUNTER: Over 200 species of birds and 23 mammals

SPECIAL OPPORTUNITIES: Fishing, night-spotting nocturnal animals

ECO-INTERPRETATION: Discussion of the impact of tourism, global warming and animals introduced to the area

RESTRICTIONS: None

REPRESENTATIVE MENU: *Breakfast*—varies according to boat, camp or hotel; pancakes, eggs, cereal, hot tea, fruit; *Lunch*—hot dishes, fresh fish, salad, spaghetti, cold cuts, bread, fruit, dessert; *Dinner*—seafood, meat, vegetables, pasta salad (vegetarian entrees available); *Alcohol*—wine and beer available

ABOUT THE COMPANY

BIOLOGICAL JOURNEYS
1696 Ocean Dr., McKinleyvile, CA 95521, (707) 839-0178
Biological Journeys contributes to local research activities and provides passengers with opportunities to donate directly to local research and conservation organizations like the Echida Research Center on Kangaroo Island. Biological Journeys also works with environmentally concerned organizations by fund-raising through the company's travel programs. A for-profit company owned by Ron LeValley and Ronn Storro-Patterson, it has been in business for 12 years and is operated by a staff of five people. Biological Journeys specializes in trips to Alaska, Baja California, Australia, the Amazon and the Galápagos.

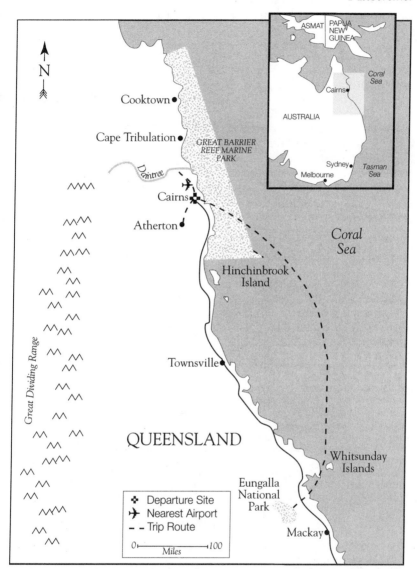

Australia

ECOFOCUS Australia's bird breeding islands and coral reefs are threatened by the impact of tourism. Global warming is adversely affecting the Great Barrier Reef. The delicate balance of the native animals has been offset by animals introduced to the area. Conservation programs are integral to the survival of the complex systems of this area.

ASMAT: WHERE THE SPIRITS DWELL
Explore the customs of Indonesian New Guinea with Zegrahm Expeditions

Arrive by boat in the Gorong Island Group and begin your island journey in the village of Amar with a presentation of a traditional knife dance and trance dance. There will also be time for snorkeling, swimming and limestone cliff exploration on the island of Manawoka. Visit Triton Bay and explore its maze of inlets and channels, swim among the colorful fish or venture into the forests in search of hornbills, cockatoos and parrots. From Timuka, an island hidden behind thick tangles of mangrove trees, you will sail into Agats, the center of the Asmat, also called Irian Jaya. You will venture up the river through narrow waterways to remote villages such as Ocenep, Pirien, Biwar Laut and Owus. Local residents will meet the group from their canoes and lead them into their village, where various traditions, such as wood sculpting, will be shared with the group. After visiting the Asmat, you will cruise to the Aru Islands in the Arafura Sea, home to flying fish and spinner dolphins. The Aru Islands are also a rich habitat for birds and other wildlife. The final four days of the trip will be spent visiting the Kur/Kaimeer, Watubela and Banda islands, where you will interact with local villagers and walk among unique vegetation and wildlife of these tropical paradises.

ABOUT THE TRIP

LENGTH OF TRIP: 17 days
DEPARTURE DATES: Sept. 25
TOTAL COST: $4,980
TERMS: Checks; deposit 25% of total cost; cancellation policy varies according to time of notification
DEPARTURE: Arrive in Denpasar, Bali, and travel from airport to hotel with the group.
AGE/FITNESS REQUIREMENTS: Good health
SPECIAL SKILLS NEEDED: Must be open-minded
TYPE OF WEATHER: Hot, humid; 80-90°
OPTIMAL CLOTHING: Lightweight clothing
GEAR PROVIDED: Snorkeling equipment
GEAR REQUIRED: None
CONCESSIONS: In Bali and Ambon but not probable along route
WHAT NOT TO BRING: Formal attire

YOU WILL ENCOUNTER: Indigenous Asmat people; marine and bird life
SPECIAL OPPORTUNITIES: Fishing, snorkeling and diving
ECO-INTERPRETATION: Discussion of the effects of oil exploration, logging and cultural deterioration
RESTRICTIONS: Group leaders will discuss dress restrictions, photography and environmental standards per specific area.
REPRESENTATIVE MENU: *Breakfast*—cereal, fruit, eggs, bread, coffee, tea, juice, yogurt; *Lunch*—Indonesian buffet, fresh fish, fried rice and fruit; *Dinner*—barbecue, traditional dishes, continental cuisine; *Alcohol*—available on ship and you can bring your own, but alcohol is not allowed on shore

ABOUT THE COMPANY

ZEGRAHM EXPEDITIONS
1414 Dexter Ave. N., #327, Seattle, WA 98109, (206) 285-4000
Zegrahm Expeditions, a for-profit company, has been in business for two years with a staff of eight. Offering expeditionary adventures worldwide, Zegrahm Expeditions is dedicated to environmentally conscious travel and the support of local economies by using local guides and accommodations. They attempt to educate local people about environmental issues while also preserving cultural traditions through projects like forest preservation, by which the economy is stimulated through tourism rather than logging.

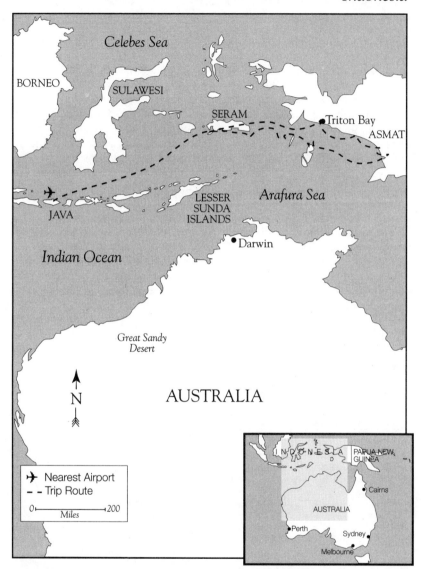

ECOFOCUS The indigenous people and wildlife of the Asmat are threatened by the potential effects of deforestation and oil exploration, including the influx of large numbers of people to a remote area and the possible leaks from pipelines. The impacts of such industries would adversely affect the traditions of the local cultures, as well as destroy the habitats of endemic birds such as the bird of paradise.

INDONESIA WILDLIFE ADVENTURE
Walk, snorkel and boat in Indonesia with Bolder Adventures

Exploring Java, Borneo, Bali, Komodo and Krakatau—this action-filled adventure will take you in search of exotic wildlife, including the nearly extinct Javan rhinoceros, the magnificent orangutan and the dinosaur-like Komodo dragon, which grows up to nine feet in length. You will travel through such diverse habitats as pristine rainforests, steaming volcanoes, virgin coral reefs and Serengeti-like plains as you hike, snorkel, swim and boat your way around Indonesia. With tours of Jakarta, Ujong Kulon National Park and Krakatoa volcano and a search for the Javan rhinoceros, the journey will challenge even the most adventurous souls. After a short trip to Bandung, Java, the group will stop at the Bogor Botanic Gardens, Tangkuban Perahu volcano and a tea plantation. In Borneo, a riverboat will take you deep into Tanjung Puting National Park, where you will canoe to the proboscis monkey research center, take jungle hikes and observe feeding at the Leakey Orangutan Research Center. You may choose to continue your journey in Bima, where a boat will take you to Komodo Island in search of Komodo dragons, with snorkeling en route. On Rinca Island, you can take a game walk and try snorkeling in a special secret cove. Share highlights of the trip with newfound friends over a farewell dinner complete with Balinese dancing before departing in the morning from Ubud.

ABOUT THE TRIP

LENGTH OF TRIP: 15-22 days
DEPARTURE DATES: June 5, July 3, 31, Sept. 25
TOTAL COST: $2,250-$3,550
TERMS: Cash, check; $300 VISA or MC deposit; cancellation policy varies according to proximity of departure date
DEPARTURE: Arrive in Jakarta; further transportation provided
AGE/FITNESS REQUIREMENTS: Able to keep up with fast-paced, activity-oriented adventure
SPECIAL SKILLS NEEDED: Sense of humor
TYPE OF WEATHER: Warm, humid climate with a chance of rain; low 68°, high 87°
OPTIMAL CLOTHING: Cool, lightweight cottons, sturdy walking shoes
GEAR PROVIDED: None
GEAR REQUIRED: Mask, snorkel, fins
CONCESSIONS: Some film and personal items are available in Jakarta or large cities in Bali; bringing these items from home is recommended.

WHAT NOT TO BRING: Excess baggage, electrical appliances
YOU WILL ENCOUNTER: Indigenous people (Balinese, Sumbanese and Javanese); exotic wildlife including Komodo dragons, orangutans and Javan rhinos
SPECIAL OPPORTUNITIES: Fishing, photography, canoeing, swimming
ECO-INTERPRETATION: A wildlife biologist and professor of zoology will lead discussions about wildlife of the area.
RESTRICTIONS: Respect of Muslim and Hindu traditions and rules
REPRESENTATIVE MENU: *Breakfast*—eggs, toast, fruit, juice, coffee, tea; *Lunch*—fried rice or noodles, vegetables with chicken; *Dinner*—seafood, chicken or beef, rice, vegetables, fresh fruit, coffee, tea; *Alcohol*—beer is available in most areas, but because of Muslim beliefs no other alcohol will be permitted

ABOUT THE COMPANY
BOLDER ADVENTURES
P.O. Box 1279, Boulder, CO 80306, (800) 397-5917, (303) 443-6789
With the aim to get travelers close to the people, cultures and natural history of Southeast Asia, Bolder Adventures is a for-profit adventure travel organization that has been in business five years with a staff of 10 people. Donations of $125 per client go toward various wildlife preservation projects in Indonesia like Save the Orangutan Foundation, and toward educating the local guides on the essentials of low-impact camping.

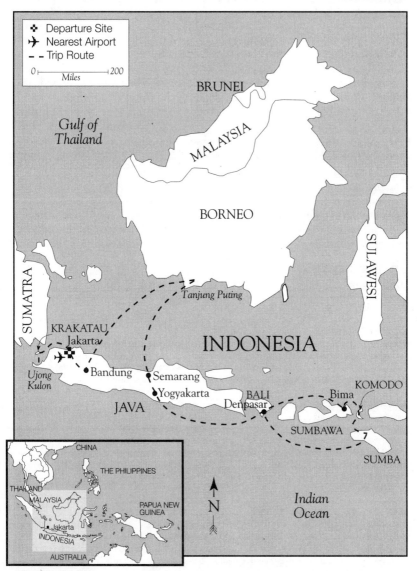

ECOFOCUS Population pressure, deforestation and trade of illegal skins are all significant threats to Indonesia's environment. Although reserves and park areas provide some environmental protection, Indonesia's natural riches are under pressure from its population of 180 million people.

ISLAND HOPPING IN THE BAY OF ISLANDS
Sea kayak in the South Pacific with New Zealand Adventures

Kayak along a shore of sandy beaches, sea caves, arches, lava gardens and hidden natural harbors. Island hopping in the Bay of Islands will take you through the 100-mile Maritime Park, with stops at aboriginal Maori archaeological sites, gorgeous beaches and old lighthouses. The route will provide the opportunity to visit all seven major islands in the bay. Explore caves, snorkel, hike, swim or kayak with dolphins through the blue waters of the Maritime Park. Several of the islands are bird sanctuaries filled with rare and endangered species such as kiwis, rails, parrots, moas, saddlebacks, black stilts, penguins, albatross and the Australian gannet. Many of these species are threatened by predatory animals introduced to the area. Intensive conservation work by New Zealand Park Services has maintained some species, such as the Chatham Islands black robin. Local preservation issues are a focal point of the trip. Thirty percent of New Zealand is protected as publicly owned national parks, forest parks and other reserves. Natural and environmental history is provided by the trip leader and owner, Mark Hutson.

ABOUT THE TRIP

LENGTH OF TRIP: 5 or 10 days
DEPARTURE DATES: Dec.-Apr.
TOTAL COST: $750-$1,195
TERMS: Cash, check; $200 deposit; refund for cancellation varies according to time of notification
DEPARTURE: Arrive at Auckland International Airport and travel independently by bus or plane to Paihia, where group is met by Mark Hutson
AGE/FITNESS REQUIREMENTS: Old enough to handle a kayak; good health
SPECIAL SKILLS NEEDED: Kayaking skills are helpful but not required.
TYPE OF WEATHER: Subtropical climate; temperatures 65°-75° with some rain showers
OPTIMAL CLOTHING: Shorts, T-shirts, paddling jacket and jogging shoes
GEAR PROVIDED: Kayak, paddle, sprayskirt, paddling jacket

GEAR REQUIRED: Sleeping bag, tent
CONCESSIONS: Film, personal items and souvenirs available in Paihia
WHAT NOT TO BRING: Winter sleeping bag or heavy, bulky items
YOU WILL ENCOUNTER: Indigenous people (Maori), marine life and scenic coastline
SPECIAL OPPORTUNITIES: Spear and reel fishing, cave exploration, swimming
ECO-INTERPRETATION: Preservation and clean-up projects are discussed.
RESTRICTIONS: None
REPRESENTATIVE MENU: *Breakfast*—fruit, cereal, eggs; *Lunch*—fruit, raw vegetables, cheese; *Dinner*—vegetables, seafood, meat, vegetarian option, fruit, dessert; *Alcohol*—may be carried personally or by guide

ABOUT THE COMPANY

NEW ZEALAND ADVENTURES
11701 Meridian North Seattle, WA 98133, (206) 364-0160
New Zealand Adventures is a for-profit company, owned by Mark Hutson, that has been in business for six years with a staff of five people. New Zealand Adventures emphasizes preservation on its trips to New Zealand and contributes to several international environmental groups. Hutson also organizes cleanups of local campsites with the assistance of local kayaking guides.

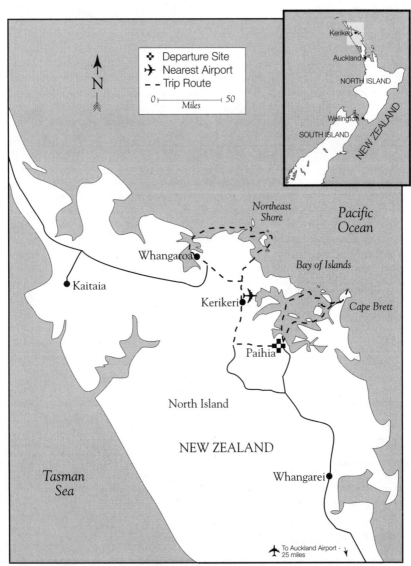

ECOFOCUS Litter caused by tourists is a major problem along the coasts of the North Island. Laws on the bird sanctuaries in the Bay of Islands attempt to enforce strict cleanup regulations. Deforestation is also a threat to the local environment; only one-third of the indigenous forest remains.

NEW ZEALAND ADVENTURE
Raft, bike, fish and hike on the south island with Wilderness River Outfitters

New Zealand's South Island offers a diverse sampling of beaches, fjords, forests, mountains and wildlife. Some 10 percent of the world's endangered birds live in New Zealand, including the black stilt, takahe, and saddleback. On this 16-day trip, you will raft, bike, hike and fish your way through the incredible scenery of South Island. Hiking in the Mt. Cook area will provide views of majestic snowcapped peaks and massive glaciers. An overnight trek to either Mueller or Hooker Hut will immerse you in the ruggedness and serenity of the Mt. Cook area. A scenic air flight will bring you to the Makarora area, a forested valley flowing with four different rivers. People interested in fishing and biking will consider Makarora a paradise. The locals will share their special fishing technique of spotting their prey first and then casting a line directly to a particular fish. Water lovers in the group will find the Landsborough River in the Makarora area, where they will spend three days rafting. The river winds through forested canyons and grassy meadows, providing an excellent opportunity for studying the flora and fauna of the area. If your adventure does not feel complete, five-day extensions are available for the Milford Track.

ABOUT THE TRIP

LENGTH OF TRIP: 16 days with a 21-day option

DEPARTURE DATE: Jan.

TOTAL COST: $2,400

TERMS: Check or money order; 25% deposit; $100 cancellation fee if notification is 60 days in advance

DEPARTURE: Arrive in Christchurch.

AGE/FITNESS REQUIREMENTS: Good physical condition

SPECIAL SKILLS NEEDED: None

TYPE OF WEATHER: Varies according to altitude, 40°-90°

OPTIMAL CLOTHING: Jacket, raingear, hiking gear, running shoes, gloves and a hat

GEAR PROVIDED: All basic camping gear is available for rental use.

GEAR REQUIRED: Tent, sleeping bag, bike helmet, wet suit, fishing gear, backpack, day pack and flashlight

CONCESSIONS: Film, souvenirs and personal items are available in Christchurch.

WHAT NOT TO BRING: Excess gear

YOU WILL ENCOUNTER: Deer, various bird species, bush forests, glaciers, beaches, country farm bed and breakfast

SPECIAL OPPORTUNITIES: Fishing in Makarora area

ECO-INTERPRETATION: Ecology and natural history are discussed in addition to the effects of tourism.

RESTRICTIONS: Camping restrictions

REPRESENTATIVE MENU: *Breakfast*—fresh fruits, eggs and cereal; *Lunch*—sandwiches with fresh meats, vegetables, fruit and snack foods; *Dinner*—pork chops, chicken, tacos, fruit, salad, vegetables, dessert

ABOUT THE COMPANY
WILDERNESS RIVER OUTFITTERS
P.O. Box 871B, Salmon, ID 83467, (208) 756-3959, (800) 252-6581

Through donations to the Nature Conservancy and various Idaho environmental groups, Wilderness River Outfitters contributes to preservation efforts in the US and abroad. In New Zealand, it makes contributions to the Albatross Colony, the Yellow-eyed Penguin Breeding Ground and the New Zealand Alpine Mountain Club. It uses local lodgings and practices low-impact camping. This for-profit company is 20 years old, owned by Joe and Fran Tonsmeire and operated by a staff of 12.

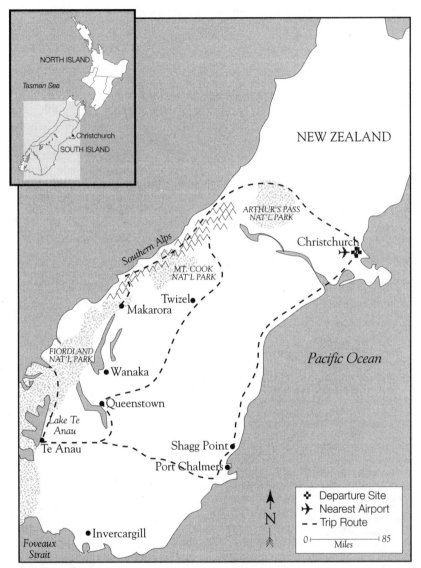

ECOFOCUS New Zealand is the home of several endemic bird species, including kiwis, flightless rails and flightless parrots. Since humans introduced mammal predators to the islands, many of the birds are threatened by extinction.

PAPUA NEW GUINEA PARADISE
Explore Papua New Guinea with International Expeditions

Begin your journey by exploring the area around the Karawari River, where you will stay in a lodge built entirely of local bush materials such as wood and thatch. Wild birds such as hornbills and black-capped lories frequent the lodge from the surrounding jungle. You will visit several nearby villages and take walks through the primary rainforest in the area, which is filled with animals like the eclectus parrot, the riflebird, the bird of paradise, the tree kangaroo and the long-nosed echidna. The *Sepik Spirit*, a touring vessel, will take you downriver to the Karawari and Sepik river basins. The Sepik River will lead you to the Chambri Lakes, a good place to watch birds and crocodiles. Along the route, you will visit several villages with differing local cultures. Stopping in Timbunke, you will fly to the land of the Huli tribe, located in the southern highlands. In the village, you can learn about their belief in ancestral spirits and sorcery. The local alpine forests are lined with waterfalls and home to 13 species of birds of paradise. After a few days, you will travel to the Central Highlands for a glimpse of the lifestyles of the Wahgi people before you fly to Cairns, Australia. Once there you will have ample opportunity to view Australian shorebirds or take a plunge in a mountain stream before the long flight home.

ABOUT THE TRIP

LENGTH OF TRIP: 17 days
DEPARTURE DATES: July, Sept., Nov.
TOTAL COST: $3,764
TERMS: Check; $300 deposit; $25 cancellation fee, which increases according to notification date
DEPARTURE: Arrive in Cairns, Australia, and take a connecting flight to Mt. Hagen, Papua New Guinea, where vans will take you to your hotel.
AGE/FITNESS REQUIREMENTS: None
SPECIAL SKILLS NEEDED: None
TYPE OF WEATHER: Warm, 80° with afternoon thunderstorms
OPTIMAL CLOTHING: Cool long-sleeve shirts, pants and walking shoes
GEAR PROVIDED: None
GEAR REQUIRED: Cameras, binoculars
CONCESSIONS: Will be available in certain locations; advised to bring with you

WHAT NOT TO BRING: Appliances or fancy clothes
YOU WILL ENCOUNTER: Indigenous people (Huli, Wahgi), brilliant plant and animal life
SPECIAL OPPORTUNITIES: Photography, swimming and some fishing
ECO-INTERPRETATION: Information about forest encroachment and cultural degradation is provided.
RESTRICTIONS: Photography of indigenous people is not always appropriate; discerning demeanor in approach of indigenous people is discussed by guides.
REPRESENTATIVE MENU: *Breakfast*—eggs, cereal, bread, fruit, coffee, tea, sausage, bacon; *Lunch*—some packed lunches, local cuisine, assorted vegetables and meat; *Dinner*—hotel menu, European fare, assorted vegetables and meat; *Alcohol*—available for purchase at lodges

ABOUT THE COMPANY

INTERNATIONAL EXPEDITIONS
One Environs Park, Helena, AL 35080, (205) 428-1700, (800) 633-4734
International Expeditions is a for-profit company directed by Richard Ryel and run by a staff of 30 for 12 years. Specializing in nature tours around the world, International Expeditions uses local land operators, guides and airlines. It contributes to the local economy and makes donations to the Amazon Center for Environment and Education Research.

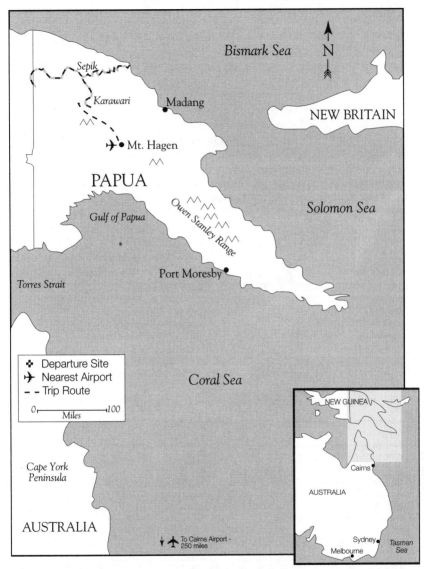

ECOFOCUS Over 70 percent of Papua New Guinea is covered in rainforest and filled with thousands of species of plants and animals. The discovery of oil, the demand for timber and agriculture, indiscriminate hunting of wildlife and the desire of the indigenous people to have Western luxuries are threatening the intricate balance of this remote area.

ISLANDS OF THE SOUTHWEST PACIFIC
Birdwatch in the Southwest Pacific with Victor Emanuel Nature Tours

Indulge in the enormous variety of bird life of the Southwest Pacific. The journey begins on the island of Upolu, in the native forests where endangered birds like the tooth-billed pigeon, purple-capped fruit dove, Pacific pigeon, white-rumped swiftlet and many other birds of the South Pacific live. After a final search for endemic species you will fly to Viti Levu, the largest of the Fiji Islands and the home of the yellow musk parrot, black-faced shrikebill and the elusive pink-billed parrot finch. The next stop on this island tour, the island of Taveuni, encompasses humid lowlands, balmy beaches and misty, moss-adorned forests. Bird species here include the Peale's pigeon, fan-tailed cuckoo and silktail and Vanikoro flycatchers. Returning to Viti Levu, you will have another day to search for species you may not have found before you fly to Noumea, the capital of New Caledonia and the base for your search for birds like the kagu and the red-crowned parakeet. Birding will focus on the Rivierre Bleu forest reserve, which is filled with unique flora and fauna. The next 12 days will be divided between Efate, Guadalcanal, Vanicoro, San Cristobal, Malaita, Honiara, Papua New Guinea, West New Britain and Australia, where you will continue the search for an incredible variety of birds.

ABOUT THE TRIP

LENGTH OF TRIP: 28 days
DEPARTURE DATES: Aug. 30
TOTAL COST: $6,175
TERMS: Cash, Visa, MC, AmEx; $300 deposit; cancellation fee of $75 or more depending on time of notification
DEPARTURE: Join group in Los Angeles and fly to Nadi, Fiji.
AGE/FITNESS REQUIREMENTS: Current vaccinations required
SPECIAL SKILLS NEEDED: None
TYPE OF WEATHER: Humid, warm climate with pleasant temperatures; 85° daytime, 70° night
OPTIMAL CLOTHING: Cool long-sleeved shirts, pants, broad-brimmed hat, hiking boots, mid-calf wool socks
GEAR PROVIDED: Spotting scope

GEAR REQUIRED: Waterproof gear, binoculars, two small duffel bags, day pack, umbrella
CONCESSIONS: Film can be purchased along trip; other items available in Nadi
WHAT NOT TO BRING: Many pieces of luggage—limited room for baggage
YOU WILL ENCOUNTER: Incredible bird life including several endangered species (kagu, tooth-billed pigeon, blue-crowned lory, simoan fantail, orange dove and silktail)
ECO-INTERPRETATION: Ornithological and natural history information provided by local guides
RESTRICTIONS: None
REPRESENTATIVE MENU: Breakfast and lunch—American-style, with fresh tropical fruit; Dinner—island cuisine: fresh seafood and fruit (American-style also available)

ABOUT THE COMPANY

VICTOR EMANUEL NATURE TOURS
2525 Wallingwood Dr., Ste. 1003, Austin, TX 78746, (512) 328-5221, (800) 328-VENT
Victor Emanuel Nature Tours, owned by Victor Emanuel, is a for-profit company that has been in business for 18 years with a staff of 17. It specializes in natural history and birding tours around the world. Through donations, conservation tours for nonprofit organizations, staff representation on various conservation boards and contributions to research efforts, Victor Emanuel Nature Tours makes a strong contribution to environmental protection.

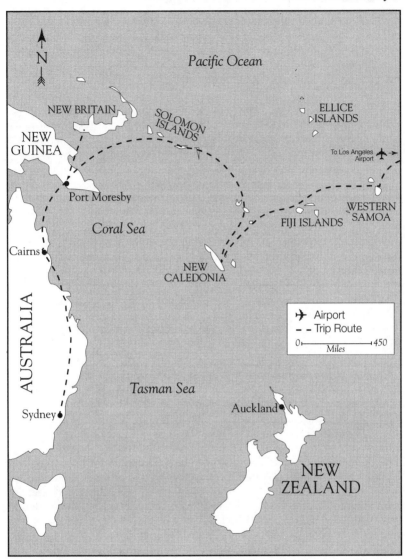

ECOFOCUS The islands of the Southwest Pacific are filled with the remaining populations of several endangered bird species that until recently were being eliminated by animals introduced to the area by human beings. Protecting these areas from the effects of insensitive tourism and conserving the fragile environments of these birds is essential to the survival of the species endemic to the Southwest Pacific.

ECOTOURISM PUBLICATIONS

A Guide to EarthTrips: Nature Travel on a Fragile Planet, (Living Planet Press, Los Angeles, 1991, $12.95) is a Conservation International Book written by Dwight Holing. Well-organized and easy to follow, this book's vivid descriptions take you to the ends of the earth. Directories of volunteer vacations, tour operators and resources in the back.

Soft Paths: How to Enjoy the Wilderness Without Harming it, (National Outdoor Leadership School, Lander, WY, 1988, $10.95). Bruce Hampton, a former NOLS instructor—one of the world's most venerable wilderness schools—has produced a solid guide to low-impact camping. Send $10.95 plus $3.50 shipping to NOLS Mail Order, P.O. Box 18, Lander WY 82520.

Pathfinder: The Insider's Guide to Adventure Travel. The first issue of this monthly newsletter for adventurous travelers is scheduled to be published this fall. It will turn an eye toward ecological impacts of travel and provide information on service trips. It will be entirely paid for by subscriptions, and will run no paid advertisements. Contact Pathfinder at P.O. Box 8656, Missoula, MT 59807, (406) 542-0449.

The **Directory of Alternative Travel Resources** lists 250 alternative and socially-responsible travel opportunities, about one-quarter of them environmental. For a copy, send $7.50 to One World Family Travel Network, 81868 Lost Valley Lane, Dexter, OR 97431.

The Green Travel Sourcebook (John Wiley & Sons, New York, 1992, $14.95), by veteran travel writers Daniel and Sally Grotta, is a richly-researched guide with chapters on everything from "Green Cruises—From Small Sloops to Stately Ships" to "Needing to be Needed—Programs for Those Who Want to Give of Themselves."

The two-volume World Wildlife Fund study, **Ecotourism: Potentials and Pitfalls,** by Elizabeth Boo offers an overview of ecotourism, and case studies on five countries: Dominica, Ecuador, Mexico, Costa Rica and Belize. For both volumes send $25.50 to World Wildlife Fund, P.O. Box 4866, Hampden Post Office, Baltimore, MD 21211.

Eco-Journeys: The World Guide to Ecologically Aware Travel and Adventure (Noble Press, Chicago, 1992, $15.95), by Stephen Foehr, is a valuable reference guide to travel and recreation, highlighting some of the most exciting trips around the world.

Tensie Whelan's book **Nature Tourism: Managing the Environment** (Island Press, Washington, D.C., 1991, $19.95) covers the ups and downs of this new field with plenty of concrete evidence from Kenya, Costa Rica and the Western United States.

Policies for Maximizing Nature Tourism's Ecological and Economic Benefits, a 38-page World Resources Institute report written by Kreg Lindberg, focuses on how tourism can generate funding for environmental protection by direct means like entrance fees, as well as lessen pressure to enter into natural areas for food and fiber. Send $12.50 plus $3 shipping to World Resources Institute, P.O. Box 4852, Hampden Station, Baltimore, MD 21211, or phone credit card orders, (800) 822-0504.

ORGANIZATIONS PROMOTING ECOTRAVEL

The Ecotourism Society is a nonprofit organization dedicated to conserving natural environments and sustaining the well-being of local people through responsible travel. Founded in 1990 to serve tourism and conservation professionals, The Ecotourism Society also provides discounts to members on a host of ecotourism publications and resources. It also publishes a quarterly newsletter and other information on visitor management, impacts, literature, gear and contacts for ecosystems worldwide. Along with the World Wildlife Fund, the Nature Conservancy, the Sierra Club and the National Audubon Society, the society is currently working to develop "green evaluations" of tour operators and lodges, planned for publication in 1994. Contact The Ecotourism Society, 801 Devon Place, Alexandria, VA 22314, (703) 549-8979.

The Center for Responsible Tourism was formed in 1984 to involve North Americans in ethical issues about travel in the third world. It advocates travel that benefits both host and guest and is harmful to neither. A nonprofit organization, the Center for Responsible Tourism produces a quarterly newsletter called "Responsible Traveling" for its contributors. For information, send a stamped, self-addressed envelope to the Center for Responsible Tourism at P.O. Box 827, San Anselmo, CA 94979, (510) 843-5506.

The Adventure Travel Society (ATS), is the only trade association in the world that addresses the issues related to nature-based tourism and adventure travel. ATS organizes the World Congress on Adventure Travel and Ecotourism. For more information, contact the society at 6551 S. Revere Pkwy., Ste. 160, Englewood, CO 80111, (303) 649-9016.

Cultural Survival, defends the rights of indigenous people and ethnic minorities on five continents. The 20-year-old organization works with native peoples around the world to help them adapt new technologies to their traditional systems of production. Cultural Survival also publishes an informative quarterly magazine. Contact Cultural Survival, Inc., 215 First St., Cambridge, MA 02142, (617) 621-3818.

Lonely Planet guides are dependably clear-eyed, fiesty, and pay plenty of attention to environmental, cultural and social issues in the numerous countries they cover. Contact Lonely Planet Publications, Embarcadero West, 112 Linden St., Oakland CA 94607.

Traveling by camel in the Sahara Desert in Morrocco.

DIRECTORY OF ECOTRAVEL OUTFITTERS

A journey of a thousand miles begins with a single phone call. Call or write for information on the trips that excite and entice you. Dream trips *can* become reality—if you take that first step, and contact an outfitter.

1. **A LAZY H OUTFITTERS**
P.O. Box 729
Choteau, MT 59422
(406) 466-5564
Types of Adventure: Wilderness horseback adventure. **Destinations:** Bob Marshall Wilderness in Montana Rocky Mountains and The Chinese Wall—a giant overthrust belt in the Rockies-Continental Divide. **Special Emphasis:** Horseback riding, blue-ribbon trout fishing, hiking to high alpine lakes, photography and wildflowers. **Average Cost:** $1,500/person for 10 days.

2. **AAT KINGS ADVENTURE TOURS**
512 S. Verdugo Dr., Ste. 200
Burbank, CA 91502
(818) 843-5905
Types of Adventure: Soft adventure camping safaris by 4-wheel-drive and coach. **Destinations:** Australia. **Special Emphasis:** Wildlife, aboriginal art and culture, national parks, Australian outback. **Average Cost:** $716-$1,941 for 6-16 days.

3. **ABOVE THE CLOUDS TREKKING**
P.O. Box 398
Worcester, MA 01602-0398
(508) 799-4499, (800) 233-4499
Types of Adventure: Culturally oriented trekking and walking trips. **Destinations:** Nepal, Bhutan, Pakistan, India, Norway, Austria, France, Great Britain, Ireland, Poland, Madagascar, Kenya, Zimbabwe, Botswana, Argentina, Costa Rica. **Special Emphasis:** Walking trips through remote mountainous areas; coming into contact with local people who are unaccustomed to outsiders; culturally sensitive. **Average Cost:** $2,000/person for land package for a 3-week trip.

4. **ACTION WHITEWATER ADVENTURES**
P.O. Box 1634
Provo, UT 84603
(800) 453-1482
Types of Adventure: Whitewater rafting trips. **Destinations:** Main Salmon River (Idaho), Middle Fork of the Salmon (Idaho), Colorado River/Grand

Canyon, American River (California). **Special Emphasis:** Quality rafting vacations. **Average Cost:** $900/person/week, $180/person/2 days.

5. **ADVENTURE ASSOCIATES**
   P.O. Box 16304
   Seattle, WA 98116
   (206) 932-8352

Types of Adventure: Off-the-beaten-path wilderness adventures for the responsible traveler. Kayak, llama trek, bike, ski, climb, hike, raft, safari, outdoor leadership. Women in Recovery Wilderness Trek, Wilderness and Spirit, Empowerment through Challenge, Personal Recovery. **Destinations:** East Africa, Morocco, Costa Rica, Baja, Copper Canyon/Mexico, Pacific Northwest. **Special Emphasis:** Women and coed trips, small groups (6-10), culturally and ecologically sensitive, expert local guides, nature education and personal discovery. **Average Cost:** $50-$150/day.

6. **ADVENTURE CENTER**
   1311 63rd St., Ste. 200
   Emeryville, CA 94608
   (510) 654-1879

Types of Adventure: Wildlife safaris, treks, overland expeditions, cycling, train journeys, cultural trips, camel trips. **Destinations:** Worldwide—over 100 countries, including Africa, Asia, Latin America, South Pacific, Europe, Middle East. **Special Emphasis:** Cultural interaction with local people, flexible itineraries, international groups, low environmental impact. **Average Cost:** $500-$7,000 depending on itinerary length. Trips from 1 week to 6 months.

7. **ADVENTURES IN PARADISE**
   155 W. 68th St., Ste. 525
   New York, NY 10023
   (212) 595-5782, (800) 736-8187

Types of Adventure: Nature trips, national parks, birdwatching. **Destinations:** Thailand, Indochina, Indonesia. **Special Emphasis:** Personalized tours, individual service. **Average Cost:** Varies.

8. **THE AFRICA ADVENTURE COMPANY**
   1620 South Federal Highway, Ste. 900
   Pompano Beach, FL 33062
   (305) 781-3933, (800) 882-9453

Types of Adventure: Wildlife safaris, photo safaris, walking safaris, canoe safaris, chimpanzee safaris, scuba diving, mountain climbing, whitewater rafting, kayaking, gorilla safaris, hiking. **Destinations:** Zimbabwe, Zambia, Botswana, Tanzania, Kenya, Rwanda, Zaire, South Africa, Malawi, Seychelles, Mauritius. **Special Emphasis:** Wildlife safaris. **Average Cost:** $3,000-$5,000/person land cost.

9. **AFRICAN ENVIRONMENTS/ MOUNTAIN MADNESS**
   4218 SW Alaska, Ste. 206
   Seattle, WA 98116

(206) 937-8389, Fax (2(

Types of Adventure: Moun and instruction, photo safari tural exploration. **Destinati** China, Nepal, Caucasus M ic Northwest. **Special E** educational, small group and cultural emphasis, personaliz tom trips. **Average Cost:** $130/day.

10. **AFRICATOURS**
    210 Post St., #911
    San Francisco, CA 94108
    (415) 391-5788

Types of Adventure: Educational, photo safaris, canoeing, rafting, hiking, mountain climbing, wildlife observations. **Destinations:** Botswana, Kenya, South Africa, Malawi, Tanzania, Zimbabwe, Zambia. **Special Emphasis:** Wildlife observation. **Average Cost:** $2,500.

11. **ALASKA DISCOVERY, INC.**
    234 Gold St.
    Juneau, AK 99801
    (907) 586-1911, Fax (907) 586-2332

Types of Adventure: Sea kayaking, canoeing and rafting. **Destinations:** Glacier Bay, Admiralty Island, southeast Alaska, Brooks Range in the Arctic National Wildlife Refuge, Tatshenshini/Alsek rivers. **Special Emphasis:** Low-impact, comfortable camping. **Average Cost:** $1,100 for 5-day tour, $1,350 for 7-day tour, $2,200 for 10-day Arctic trip, $1,700 for 10-day Tatshenshini raft trip.

12. **ALASKA UP CLOSE**
    P.O. Box 32666
    Juneau, AK 99803
    (907) 789-9544

Types of Adventure: Travel in small groups by charter boat, mini-van, light aircraft, day hikes—all with interpretive guides. **Destinations:** Alaska, featuring southeast Alaska's Inside Passage and optional extensions to all parts of the state. **Special Emphasis:** Nature photography, birding, natural history, wildlife viewing, custom itineraries for independent travelers and small groups, specialty travel for nonprofit organizations. **Average Cost:** $1,500-$1,850 for 7-day package.

13. **ALASKA WILDLAND ADVENTURES**
    P.O. Box 389
    Girdwood, AK 99587
    (907) 783-2928, (800) 334-8730

Types of Adventure: Alaska natural history safaris, senior safaris, wilderness expeditions. **Destinations:** Denali National Park, Kenai Fjords National Park, Kenai National Wildlife Refuge, Chugach National Forest, seacoast glaciers, Inside Passage. **Special Emphasis:** Natural history tours that take travelers beyond the conventional bus tours and cruises. Trips are informative and experience-oriented with

of 18. **Average Cost:** $1,950 for 7 days, or 10 days, $2,795 for 12 days.

ALASKAN WILDERNESS SAILING SAFARIS
P.O. Box 1313
Valdez, AK 99686
(907) 835-5175, Fax (907) 835-5175
Types of Adventure: Sailing, sea kayaking. **Destinations:** Alaska (Prince William Sound). **Special Emphasis:** Educational trips on effects of *Exxon Valdez* oil spill, and natural history tours. **Average Cost:** $340-$1,420.

15. ALL ADVENTURE TRAVEL, INC.
P.O. Box 4307
Boulder, CO 80306
(303) 499-1981
Types of Adventure: Biking, hiking, nature safaris, expedition cruises. **Destinations:** US, Canada, Europe, South America, Asia, Africa. **Special Emphasis:** Soft adventure, cultural interaction. **Average Cost:** $150-$250/day.

16. AMAZON EXPLORERS, INC.
499 Ernston Rd.
Parlin, NJ 08859
(800) 631-5650, Fax (908) 721-2404
Types of Adventure: Soft adventure, photography, safaris, expedition cruises. **Destinations:** South America, Africa. **Special Emphasis:** Ecological safaris. **Average Cost:** $1,900.

17. AMAZON TOURS & CRUISES
8700 W. Flagler, Ste. 190
Miami, FL 33174
(305) 227-2266, (800) 423-2791
Types of Adventure: Amazon river trips—from expeditions to air-conditioned ships in the upper and mid-Amazon areas. **Destinations:** Iquitos, Peru; Leticia, Colombia; Tabatinga and Manaus, Brazil. **Special Emphasis:** FIT and small groups. **Average Cost:** $70/day.

18. AMAZONIA EXPEDITIONS
1824 NW 102nd Way
Gainesville, FL 32606
(904) 332-4051
Types of Adventure: Expeditions in Amazon rainforest. **Destinations:** Amazon rainforest. **Special Emphasis:** Custom itineraries for personal interests and needs. **Average Cost:** $1,375/person for 2 weeks.

19. AMERICAN WILDERNESS EXPERIENCE, INC. (AWE!)
P.O. Box 1486
Boulder, CO 80306
(303) 444-2622, (800) 444-0099
Types of Adventure: Horse packing, llama trekking, whitewater rafting, backpacking, canoeing,

sailing, mountain biking, sea kayaking, natural history, trekking, snorkeling and diving. **Destinations:** Rocky Mountain West, desert Southwest, Alaska, Hawaii, Canada, Mexico, New England, Florida, Virgin Islands, Australia, New Zealand, Peru and Belize. **Special Emphasis:** Small group, ecologically sensitive backcountry travel. **Average Cost:** $74-$175/person for 6 days, domestic; $100-$175/person for 12 days, international.

20. ANCIENT FOREST ADVENTURES
Central Oregon Environmental Center
16 NW Kansas Ave.
Bend, OR 97701
(503) 385-8633, (800) 551-1043
Types of Adventure: Interpretive trips into the ancient forests of Oregon by foot, snowshoe and cross-country skiing. **Destinations:** Ancient or old growth forests in Oregon; from the coast to the dry east side of the state. **Special Emphasis:** Ecological management. Discovering the magic and the mystery of the ancient forests through Native American lore, individual quiet time and night walks. **Average Cost:** $525/week for resort-based trips, $350/week for camping.

21. ANDES TREKKING
12021 Wilshire Blvd.
Los Angeles, CA 90025
(310) 575-3990
Types of Adventure: Trekking, wilderness training, native cultures, ecological work. **Destinations:** Chile, Patagonia. Andean temperate rainforest, Andean desert. **Special Emphasis:** Educational, environmental problems and fieldwork, wilderness and mountaineering training, native cultures. **Average Cost:** $2,800-$3,200.

22. APPALACHIAN MOUNTAIN CLUB
P.O. Box 298-ET
Gorham, NH 03581
(603) 466-2721
Types of Adventure: Guided and unguided hikes, outdoor skills workshops, nature and wildlife exploration, environmental education programs for all ages. **Destinations:** White Mountain National Forest (New Hampshire), Berkshires (Massachusetts), Catskills (New York). **Special Emphasis:** Promoting the protection, enjoyment and wise use of our natural resources. **Average Cost:** $200.

23. ARIZONA RAFT ADVENTURES
4050 ET East Huntington Dr.
Flagstaff, AZ 86004
(800) 786-7238, Fax (602) 526-8246
Types of Adventure: Minimal-impact adventure travel and whitewater rafting. **Destinations:** Arizona, Colorado. **Special Emphasis:** Preservation of Grand Canyon and rivers of western US. **Average Cost:** $980-$1,870.

24. **ASIAN PACIFIC ADVENTURES**
826 S. Sierra Bonita Ave.
Los Angeles, CA 90036
(213) 935-3156, (800) 825-1680
Types of Adventure: Bicycling, in-depth cultural and environmental trips, hiking, jungle trekking, mountain biking, overland expeditions, photography, rafting, safaris, wildlife. **Destinations:** Bali, Borneo, China, Himalayas, India, Indonesia, Japan, Malaysia, Nepal, Pakistan, Southeast Asia, Thailand, Tibet, Vietnam, Laos and Cambodia. **Special Emphasis:** Environmental and cultural adventures to spectacular, remote areas in small groups that allow in-depth experiences with minimal impact on the environment and indigenous cultures. **Average Cost:** $150/person/day.

25. **BAIKAL REFLECTIONS, INC.**
13 Ridge Rd.
Fairfax, CA 94930
(415) 455-0155, (800) 927-2797
Types of Adventure: Ocean kayaking, backpacking, cycling, wildlife, sailing. **Destinations:** Lake Baikal and other parts of the Commonwealth of Independent States (former Soviet Union). **Special Emphasis:** Educational, ethics-based, environment-centered, indigenous cultures. **Average Cost:** $1,700 land cost.

26. **BAJA EXPEDITIONS**
2625 Garnet Ave.
San Diego, CA 92109
(619) 581-3311, (800) 843-6967
Fax (619) 581-6542
Types of Adventure: Whale watching, sea kayaking, Sea of Cortez cruises, mountain biking, wilderness sailing, scuba diving. **Destinations:** Baja, Mexico and Costa Rica. **Special Emphasis:** Natural history and adventure travel. **Average Cost:** $150/day (includes air).

27. **BEAR TRACK OUTFITTING COMPANY**
P.O. Box 937
Grand Marais, MN 55604
(218) 387-1162
Types of Adventure: Canoe trips, backpacking, cross-country skiing, hiking, birdwatching. **Destinations:** Boundary Waters Canoe Area, Quetico Provincial Park, Isle Royale National Park, Superior National Forest. **Special Emphasis:** High-tech equipment including ultralight Kevlar canoes and solo canoes, entry-level guided trips geared for women, guided by women. **Average Cost:** $240 for 5-day self-guided trip that includes food and equipment; $425/5-day trip with a guide, food and equipment.

28. **BHUTAN TRAVEL, INC.**
120 E. 56th St., #1430
New York, NY 10022
(800) 950-9908, Fax (212) 750-1269

Types of Adventure: Adventure trekking and cultural exploration. **Destinations:** Bhutan. **Special Emphasis:** Small group, low-impact camping that supports indigenous conservation issues. **Average cost:** $1,550-$4,500.

29. **BICYCLE AFRICA**
4887-B Columbia Dr. S.
Seattle, WA 98108
(206) 628-9314
Types of Adventure: Culturally sensitive, environmentally friendly people-to-people bicycle tours. Easy to moderate difficulty, with extra side trips for the active. **Destinations:** Tunisia, Zimbabwe, Botswana, Kenya, Benin, Togo, Ghana, Mali, Senegal, Gambia. **Special Emphasis:** Small group bicycle tours to untouristed areas promoting cultural and environmental understanding while benefiting local economy. **Average Cost:** $990-$1,290 plus airfare, for 2 weeks.

30. **BICYCLE ROMANTIC SCOTLAND, PETER COSTELLO, LTD.**
P.O. Box 23490
Baltimore, MD 21203
(410) 783-1229, Fax (410) 539-3250
Types of Adventure: Van-supported, 6-day bicycle tours. Hiking and trekking. **Destinations:** Scotland—Sir Walter Scott's fictionalized historic Scottish "Borders." **Special Emphasis:** Challenging, varied trips, passionate countryside, glorious castles, stately houses, charming market towns, quiet rivers, uncluttered byways. **Average Cost:** $1,050.

31. **BIKECENTENNIAL, INC.**
P.O. Box 8308, 150 E. Pine
Missoula, MT 59807
(406) 721-1776
Types of Adventure: Bicycle tours, adventure and educational. **Destinations:** 30 states and provinces. **Special Emphasis:** Self-contained camping. **Average Cost:** $1,000.

32. **BILL DVORAK KAYAK & RAFTING EXPEDITIONS, INC.**
17921-Z, US Highway 285
Nathrop, CO 81236
(719) 539-6851, (800) 824-3795
Types of Adventure: Whitewater rafting, kayaking, canoeing, custom fishing, horseback riding, mountain biking, instructional seminars, classical music journeys, international expeditions. **Destinations:** Colorado, Utah, Wyoming, Arizona, New Mexico, Texas, Mexico, Hawaii, New Zealand, Australia. **Special Emphasis:** Dvorak Expeditions fosters a participatory spirit in whitewater expeditions as well as supporting river conservation and a socially responsible, low-impact style of travel. **Average Cost:** $32-$81 for half-day to full-day trips, $190-$1,929 for 2-day to 12-day trips.

33. **BILL THOMAS'S TOUCH OF SUCCESS PHOTO SEMINARS**
P.O. Box 194
Lowell, FL 32663
(904) 867-0463
Types of Adventure: Workshop study for serious photography students. **Destinations:** Africa, Alaska, Australia, upper Amazon, Southwest Indian country, Everglades, Big Cypress Swamp, Okefenokee Swamp, Great Smoky Mountains, Maine. **Special Emphasis:** Nature and wildlife photography with ecological and environmental awareness. **Average Cost:** $895-$4,000.

34. **BIOLOGICAL JOURNEYS**
1696 Ocean Dr.
McKinleyville, CA 95521
(707) 839-0178, Fax (707) 839-4656
Types of Adventure: Small group, educational adventure travel. **Destinations:** Alaska, Baja California, Australia, the Amazon and the Galápagos. **Special Emphasis:** Endangered wildlife, threatened habitats and conservation activism. **Average Cost:** $1,295-$5,300.

35. **BOLDER ADVENTURES, INC.**
P.O. Box 1279
Boulder, CO 80306
(303) 443-6789
Types of Adventure: Natural history and in-depth cultural programs, both small group and independent travel arrangements. **Destinations:** Southeast Asia, including Thailand, Indonesia, Laos, Cambodia, Vietnam, Malaysia, Burma. **Special Emphasis:** Activity-oriented programs that get you close to the people and natural beauty of Southeast Asia. **Average Cost:** $1,500-$3,000.

36. **BORTON OVERSEAS**
5516 Lyndale Ave. S.
Minneapolis, MN 55419
(612) 824-4415, (800) 843-0602
Types of Adventure: African safaris, Kilimanjaro climb, glacier and mountain hiking, rafting, whale watching, skiing, birdwatching, dog and reindeer sledding. **Destinations:** Africa, Scandinavia, Lapland, Greenland, Spitsbergen. **Special Emphasis:** Environmentally and culturally sensitive trips. Individual and small groups. **Average Cost:** $400-$4,000.

37. **BOX K RANCH**
Box 110, Buffalo Valley Rd.
Moran, WY 83013
(307) 543-2407
Types of Adventure: Backcountry horse-pack trips. **Destinations:** Teton wilderness, Wyoming adjacent to Yellowstone National Park. **Special Emphasis:** Wildlife viewing, fishing, mountain scenery. **Average Cost:** $990 for 6 days.

38. **CAL NATURE TOURS, INC.**
7310 S.V.L. Box
Victorville, CA 92392
(619) 241-2322
Types of Adventure: Ecotours, treks. **Destinations:** Southwestern US, Northern Europe, Jamaica, Eastern Europe. **Special Emphasis:** Alternative travel to promote appreciation of global environmental crises and encourage local conservation of resources. **Average Cost:** $395-$3,295.

39. **CANADIAN RECREATIONAL CANOE ASSOCIATION**
1029 Hyde Park Rd., Ste. 5
Hyde Park, ONT, N0M 1Z0, Canada
(519) 641-1261
Types of Adventure: Wilderness tourism: canoe and kayak trips. **Destinations:** Canada. **Special Emphasis:** The national volunteer nonprofit promoting safety, heritage and environmental ethics pertaining to canoeing and kayaking in Canada. **Average Cost:** $435-$2,500 (Canadian).

40. **CANADIAN RIVER EXPEDITIONS**
3524 W. 16th Ave.
Vancouver, BC, V6R 3C1, Canada
(604) 738-4449
Types of Adventure: Wilderness expeditions to remote regions in western Canada and Alaska. Natural history exploration of rivers and coastlines. **Destinations:** Tatshenshini and Alsek rivers, British Columbia and Alaska. Firth River, Chilko River, Chilcotin River and Fraser River. **Special Emphasis:** Wilderness and natural history. Small groups with expert naturalists. Each trip focuses on a different landscape and ecosystem. **Average Cost:** $1,200 for 12 days, fully guided, all-inclusive.

41. **CANYONLANDS FIELD INSTITUTE**
P.O. Box 68
Moab, UT 84532
(801) 259-7750
Types of Adventure: Ecology studies collected on various wilderness expeditions. **Destinations:** Utah, Colorado, Arizona, New Mexico. **Special Emphasis:** Monitoring the flora and fauna of the western US. **Average Cost:** $40-$750.

42. **CASCADE RAFT COMPANY**
P.O. Box 6
Garden Valley, ID 83622
(800) 292-7238
Types of Adventure: River trips and river treks. **Destinations:** Payette River and Main Salmon River (Idaho), Nepal, Costa Rica. **Special Emphasis:** Environmentally sensitive river trips and treks. **Average Cost:** $80-$120/day.

43. **CHEESEMAN'S ECOLOGY SAFARIS**
20800 Kittredge Rd.

Saratoga, CA 95070
(408) 741-5330, (408) 867-1371
Types of Adventure: Natural history, environmental sensitivity, bird and mammal observation, wildlife photography, educational. **Destinations:** East Africa (Kenya and Tanzania), Galápagos, Costa Rica, South America, Alaska, India, Botswana, Rwanda, Arctic. **Special Emphasis:** Educational trips with small groups, usually 4 participants per leader. Habitat protection, flora and fauna sensitivity. **Average Cost:** $3,300 for 18-day trip to Kenya, $2,700 for 18-day trip to Costa Rica.

44. **CLEARWATER CANOE OUTFITTERS & LODGE**
355 Gunflint Trail
Grand Marais, MN 55604
(218) 388-2254, (800) 527-0554
Types of Adventure: Wilderness canoe/camping trips or historic lodge/B&B, secluded cabins. Hiking, mountain biking, guided kayak tours on Lake Superior. **Destinations:** Minnesota's Boundary Waters Canoe Area, Canada's Quetico Park in Ontario, Isle Royale on Lake Superior. **Special Emphasis:** Quality equipment and Kevlar canoes, vegetarian menu options, personalized routing and unique sailboat camping package. **Average Cost:** $40-$115/day.

45. **COLORADO OUTWARD BOUND SCHOOL**
945 Pennsylvania St.
Denver, CO 80203-3198
(303) 837-0880, Fax (303) 831-6987
Types of Adventure: Low-impact camping, hiking, skiing and climbing. **Destinations:** Western US. **Special Emphasis:** Natural history, archaeology, geology, flora and fauna are discussed. **Average Cost:** $75-$125 a day.

46. **COLORADO RIVER & TRAIL EXPEDITIONS, INC.**
P.O. Box 57575
Salt Lake City, UT 84157
(801) 261-1789
Types of Adventure: River rafting expeditions with ecology and natural history emphasis on the Green and Colorado rivers. **Destinations:** Green River, Colorado River, Scott Matheson Wetlands Preserve. **Special Emphasis:** Field studies to determine food chains, wildlife populations, environmental assessments, environmental ethics and philosophy. **Average Cost:** $50-$750.

47. **CO-OP AMERICA TRAVEL-LINKS**
14 Arrow St.
Cambridge, MA 02138
(617) 628-2667
Types of Adventure: Personally designed tours for individuals and small groups. **Destinations:** Caribbean, Central America, Africa, China, Hawaii.

Special Emphasis: How to be a socially sible traveler. **Average Cost:** $1,200.

48. **COSTA RICA EXPERTS, INC.**
3540 NW 13th St.
Gainesville, FL 32609
(800) 858-0999, (904) 377-7111
Types of Adventure: Natural history adventure, birding, botany. **Destinations:** Costa Rica. **Special Emphasis:** Rainforests, birds, botany. **Average Cost:** $406-$1,875.

49. **COUNTRY WALKERS**
P.O. Box 180
Waterbury, VT 05676
(802) 244-1387
Types of Adventure: Walking vacations. **Destinations:** US, New Zealand, Ireland. **Special Emphasis:** Environmental preservation and natural history tours led by local specialists. **Average Cost:** $299-$2,499.

50. **CREATIVE ADVENTURE CLUB**
3007 Royce Lane
Costa Mesa, CA 92626
(714) 545-5888
Types of Adventure: Diving, rainforest, primitive tribes, trekking, caving, jungle river safaris, rafting. **Destinations:** Malaysia, Indonesia, Thailand, Philippines, Laos, Cambodia, Vietnam, Borneo, Nepal. **Special Emphasis:** Face-to-face experiences with primitive tribes, rainforest. **Average Cost:** $3,000 for 15 days.

51. **EARTH TOURS LTD.**
93 Bedford, Box 3C
New York, NY 10014
(212) 675-6515
Types of Adventure: Hiking tours, snorkeling, diving, birdwatching. **Destinations:** Dominica, West Indies, Caribbean. **Special Emphasis:** Rainforest hiking, ecotourism, naturalism, fun. **Average Cost:** $1,380, inclusive except for airfare.

52. **EARTH VENTURES**
3700 Trenton Rd.
Raleigh, NC 27607
(919) 833-6067, Fax (919) VENTURE
Types of Adventure: All types of adventure through environmentally responsible tour operators. Earth Ventures uses specific travel ethics for choosing ecotourism operators. **Destinations:** Latin America, North, Central and South America, the Caribbean, Costa Rica, Galápagos. **Special Emphasis:** True ecotourism, wilderness adventure, cultural awareness, natural history, locally supportive, rewarding experiences. A portion of profits go to approved cultural and environmental preservation groups. **Average Cost:** Varies.

53. **EARTHWATCH**
680 Mt. Auburn St., Box 403

02272
Fax (617) 926-8532
Scientific expeditions. **Des-**
. **Special Emphasis:** Envi-
monitoring global change,
jered habitats and species,
)stering world health and
.......cooperation. **Average Cost:** $1,400.

54. **ECHO: THE WILDERNESS COMPANY**
6529 Telegraph Ave.
Oakland, CA 94609
(510) 652-1600, Fax (507) 652-3987
Types of Adventure: River adventures in the
western US. **Destinations:** California, Idaho and
Oregon. **Special Emphasis:** To provide environmentally conscious rafting trips that promote a
greater appreciation of wilderness and wild rivers.
Average Cost: $80-$160 a day.

55. **ECO-TOURS DE PANAMA S.A.**
Apartado 465
Panama City 9A, Panama
(507) 363-575, Fax (507) 363-550
Types of Adventure: Nature-oriented tours. **Des-**
tinations: Panama's tropical forests. **Special**
Emphasis: The conservation of Panama's natu-
ral resources. **Average Cost:** $2,500.

56. **ECOSUMMER EXPEDITIONS**
1516 Duranleau St.
Vancouver, BC, V6H 3S4, Canada
(604) 669-7741
Types of Adventure: Sea kayaking, trekking, rafting, canoeing, photo/nature expeditions. **Destina-**
tions: Baja, Belize, Bahamas, British Columbia
Pacific coast, Yukon, High Arctic, Costa Rica. **Spe-**
cial Emphasis: Keeping the "adventure" in wilderness travel with a high level of cultural and historical
interpretation. **Average Cost:** $2,000 for 10 days.

57. **ECOTOUR EXPEDITIONS**
P.O. Box 1066
Cambridge, MA 02238
(617) 876-5817, (800) 688-1822
Types of Adventure: Natural history expeditions
to remote wilderness areas guided by highly qualified scientists. **Destinations:** Brazil, Ecuador,
Panama. **Special Emphasis:** Ecology—the inter-
relationship between living things and their natural
environment. **Average Cost:** $1,600-$2,000.

58. **ECOTOURS OF HAWAII**
P.O. Box 2193
Kamuela, HI 96743
(808) 885-7759
Types of Adventure: Hiking, bicycling, kayaking
and sailing. **Destinations:** Hawaiian islands. **Spe-**
cial Emphasis: Low-impact adventures with secu-
rity, education and appreciation of natural beauty.
Average Cost: $45-$150/day.

59. **THE EDUCATED TRAVELER**
NEWSLETTER
P.O. Box 220822
Chantilly, VA 22022
(703) 471-1063, (800) 648-5168
Types of Adventure: Education, cultural trips,
special interest, natural history, ecotours, research
expeditions, archaeology. **Destinations:** Worldwide. **Special Emphasis:** Newsletter reporting on
lesser-known, special interest and educational
travel opportunities, including specialty tour guides.
Exclusive "Directory of Museum Sponsored Tours"
available to subscribers only. Information comes
from a worldwide network of many smaller,
entrepreneurial tour companies, tour guides and
undiscovered destinations. **Average Cost:** Trips,
$2,000 and up. Subscriptions to newsletter,
$65/year, $6/preview issue.

60. **F & H TRAVEL CONSULTING**
2441 Janin Way
Solvang, CA 93463
(805) 688-2441
Types of Adventure: Rainforest lodge trips, jeep
safaris, birdwatching, fishing, trans-Pantanal tour.
Destinations: Amazon and Pantanal, Brazil.
Special Emphasis: Ecological tours of Brazil.
Average Cost: $4,000.

61. **FANTASY ADVENTURES OF EARTH,**
INC.
P.O. Box 368
Lincolndale, NY 10540
(914) 248-5107
Types of Adventure: Wilderness adventures,
including horseback riding, mountain biking, high
mountain treks, whitewater challenges, sailing and
scuba diving. **Destinations:** US, Canada, Peru,
New Zealand, Belize, Alaska and Hawaii. **Special**
Emphasis: Personal challenges, custom trips,
team building, nature photography, family and single trips. **Average Cost:** $75-$125/day, wilderness trips; $90-$175/day, dude ranch program.

62. **FISHING AND FLYING**
P.O. Box 2349
Cordova, AK 99574
(907) 424-3324
Types of Adventure: Hiking, fishing, rafting,
mountaineering, sightseeing. **Destinations:** Alaska's Gulf Coast, Wrangell-St. Elias National Park,
Glacier Bay National Park, Tongass and Chugach
national forests. **Special Emphasis:** Wildlife viewing, natural wonders, off-the-beaten-track. **Aver-**
age Cost: Varies.

63. **FORCE 10 EXPEDITIONS LTD.**
P.O. Box 34354
Pensacola, FL 32507
(904) 492-6661, (800) 922-1491
Types of Adventure: Running-oriented trips with

limited space for non-runners. **Destinations:** India, Sikkim, Nepal. **Special Emphasis:** Running, cultural, monasteries, Himalayan history. **Average Cost:** $2,000/India, $1,000/Nepal.

64. **FOUNDATION FOR FIELD RESEARCH**
P.O. Box 2010
Alpine, CA 91903
(619) 445-9264
Types of Adventure: Research expeditions. **Destinations:** Europe, West Africa, Caribbean, Mexico, United States. **Special Emphasis:** Funding field research through volunteer contributions of both time and money. **Average Cost:** $495-$1,672.

65. **FOUR CORNERS SCHOOL OF OUTDOOR EDUCATION**
East Route
Monticello, UT 84535
(801) 587-2156
Types of Adventure: Rafting, backpacking, cross-country skiing, hiking. **Destinations:** Southwestern US—Utah, Arizona, Colorado and New Mexico. **Special Emphasis:** Natural and human history of the Colorado Plateau, wilderness preservation. **Average Cost:** $500-$800.

66. **FOX GLOVE SAFARIS**
15 W. 26th St.
New York, NY 10010
(212) 545-8252, (800) 437-4807
Types of Adventure: Wildlife safaris. **Destinations:** Africa—Kenya and Tanzania. **Special Emphasis:** Conservation of the environment and wildlife, the Kenyan and Tanzanian people and tribes, their culture and colorful history, special-interest groups including family safaris. **Average Cost:** $3,800 including airfare from New York.

67. **FRANGIPANI TOURS**
5855 Green Valley Circle, Ste. 109
Culver City, CA 90230
(310) 670-7184
Types of Adventure: Customized itinerary based on customer's needs. **Destinations:** South Australia, Western Australia, South Pacific Islands, Indonesia, Asia. **Special Emphasis:** Endangered species. **Average Cost:** $2,499.

68. **GALAPAGOS NETWORK**
7200 Corporate Center Dr., Ste. 404
Miami, FL 33126-7972
(305) 592-2294, (800) 633-7972
Types of Adventure: Yacht cruises in the Galápagos Islands aboard a 20-passenger, first-class yacht, including crew and naturalist guide. **Destinations:** Ecuador and the Galápagos Islands. **Special Emphasis:** Wildlife and natural history tours, birdwatching, yacht charters (motor and sail), environmental education. **Average Cost:** $600/3 nights, $800/4 nights, $1,400/7 nights.

69. **GEO EXPEDITIONS**
P.O. Box 3656
Sonora, CA 95370
(800) 351-5041
Types of Adventure: Tented safaris, walking safaris in Africa, yacht cruises in the Galápagos Islands. **Destinations:** Kenya, Tanzania, Rwanda, Zaire, Botswana, Madagascar, Seychelles Islands, Galápagos Islands. **Special Emphasis:** Natural history tours to off-the-beaten-path areas, geared to small groups and led by qualified naturalist guides. **Average Cost:** $3,400-$7,500.

70. **GLACIER BAY SEA KAYAKING**
Box 26
Gustavus, AK 99826
(907) 697-2257
Types of Adventure: Sea kayaking, camping. **Destinations:** Alaska—Glacier Bay National Park. **Special Emphasis:** Sea kayak rentals and transportation to remote areas of Glacier Bay. **Average Cost:** $500/3 days, 2 people.

71. **GREAT EXPEDITIONS INC.**
5915 West Blvd.
Vancouver, BC, V6M 3X1, Canada
(604) 263-1505, (604) 263-1476
Types of Adventure: Natural history and soft adventure tours. **Destinations:** Zimbabwe, Botswana, Galápagos, Costa Rica, Indonesia, Thailand, Nepal, British Columbia, Yukon, Alaska. **Special Emphasis:** Natural history of the various areas being visited, the flora and fauna and how they are being affected by exploitation of the natural resources. **Average Cost:** $1,000-$6,500.

72. **GREAT PLAINS WILDLIFE INSTITUTE**
P.O. Box 7580
Jackson, WY 83001
(307) 733-2623
Types of Adventure: North American wildlife safaris with groups of 6 people. **Destinations:** Northwestern Wyoming (the Yellowstone ecosystem). **Special Emphasis:** Wildlife and scenery. **Average Cost:** $1,635/person.

73. **HAWAIIAN HEART OF THE JUNGLE JOURNEYS**
P.O. Box 1567
Makawao, Maui, HI 96768
Attn: Merlyn Sikirdji
(808) 572-5083
Types of Adventure: Day hikes, backpacking and wilderness journeys to pools, waterfalls, valleys, rainforests, beaches, lava tubes, and volcanic areas. **Destinations:** Off the beaten path areas of the Hawaiin Islands. **Special Emphasis:** Individuals to large groups journeying deep into natural environments for appreciation of flora and fauna, Hawaiian spirituality; vision questing; or simple

walks in nature. Special children's tours can be designed. **Average Cost:** $50-$200/day.

74. **HIGH ANGLE ADVENTURES, INC.**
5 River Rd.
New Paltz, NY 12561
(800) 777-CLIMB

Types of Adventure: Rock climbing school and guide service. **Destinations:** The Shawangunk Mountains in New Paltz, New York. **Special Emphasis:** Private and small group instruction (limit 3 students per instructor) with emphasis on safety and technique. Individual attention. **Average Cost:** $100-$150/day.

75. **HIGH DESERT ADVENTURES**
757 E. South Temple, Ste. 201
Salt Lake City, UT 84102
(800) 345-RAFT, (801) 355-5444

Types of Adventure: Rafting, mountain biking, hiking, combination trips with two or more of these activities. **Destinations:** Grand Canyon, Canyonlands, Dinosaur National Park, southwestern Colorado, River of No Return Wilderness in Idaho. **Special Emphasis:** Environmentally sensitive and personally challenging adventures in Western wilderness areas. **Average Cost:** $230-$1,818.

76. **HIMALAYA TREKKING & WILDERNESS EXPEDITIONS**
1900 Eighth St.
Berkeley, CA 94710
(510) 540-8040, (800) 777-TREK

Types of Adventure: Trekking, river rafting, river cruises in Siberia, safaris, cultural touring. **Destinations:** India, Nepal, Tibet, Pakistan, China (Silk Road), Bhutan, Sikkim, Siberia, Ladakh. **Special Emphasis:** Low-impact, small group sizes (6 to 8), ecological awareness information, sensitivity to conditions along route. **Average Cost:** $2,000 land cost.

77. **HIMALAYAN HIGH TREKS**
241 Dolores St.
San Francisco, CA 94103
(415) 861-2391

Types of Adventure: Socially responsible and environmentally sound Himalayan treks with a strong cultural emphasis. **Destinations:** The Nepal and Indian Himalayas and the Tibetan Plateau (Zanskar and Ladakh). **Special Emphasis:** Very small groups of 4-8 people offer greater interaction with local people. **Average Cost:** $55-$75/day plus airfare of $1,300-$1,500.

78. **HIMALAYAN JOURNEYS**
757 E. South Temple, Ste. 201
Salt Lake City, UT 84102
(800) 345-7238, (801) 355-5444

Types of Adventure: Trekking and overland journeys in the Himalayan region. **Destinations:** Nepal, Ladakh, Tibet, Bhutan and Sikkim. **Special Emphasis:** Trekking and overland journeys in the Himalayas, with insight into the mountains' natural environment and the human culture of the region. **Average Cost:** $1,850.

79. **HIMALAYAN TRAVEL, INC.**
112 Prospect St.
Stamford, CT 06901
(800) 225-2380

Types of Adventure: Trekking, hiking, wildlife safaris, jungle expeditions, natural history cruises. **Destinations:** Russia, Peru, Galápagos, Nepal, India, Tibet, Bhutan, Pakistan, Thailand, Kenya, Tanzania, Rwanda, Egypt, Morocco, Israel, Turkey, Switzerland, France, Spain, Greece, Britain, Eastern Europe. **Special Emphasis:** Trekking in Nepal, camping safaris in East Africa, hiking in Europe. **Average Cost:** $500/week plus airfare.

80. **INNERASIA EXPEDITIONS**
2627 Lombard St.
San Francisco, CA 94123
(415) 922-0448, (800) 777-8183

Types of Adventure: Travel for environmental nonprofit organizations (e.g., Nature Conservancy, Audubon), commercial treks and overland tours. **Destinations:** Asia, Europe, South America and the Pacific. **Special Emphasis:** Educational travel. **Average Cost:** $2,500.

81. **INTERLOCKEN**
RR. 2, Box 165
Hillsboro, NH 03244
(603) 478-3166, Fax (603) 478-5260

Types of Adventure: Community service adventure tours. **Destinations:** US, Asia, Africa and Latin America. **Special Emphasis:** Educational travel. **Average Cost:** $2,285-$3,595.

82. **INTERNATIONAL BICYCLE FUND**
4887-B Columbia Dr. S.
Seattle, WA 98108-1919
(206) 628-9314, Fax (206) 628-9314

Types of Adventure: Small group, culturally sensitive, environmentally friendly, people-to-people tours. **Destinations:** Africa. **Special Emphasis:** Bicycling. **Average Cost:** Varies.

83. **INTERNATIONAL EXPEDITIONS, INC.**
One Environs Park
Helena, AL 35080
(800) 633-4734

Types of Adventure: Natural history travel to 30 destinations worldwide, rainforest workshops in Costa Rica and the Amazon. **Destinations:** Amazon, Belize, Costa Rica, Galápagos, Panama, Venezuela, Hawaii, Alaska, Kenya, Tanzania, India, Malaysia, Australia, China. **Special Emphasis:** Rainforests. **Average Cost:** $2,500.

84. **INTERNATIONAL OCEANOGRAPHIC FOUNDATION**
4600 Rickenbacker Causeway
Miami, FL 33149
(305) 361-4697
Types of Adventure: Whale watching, natural history, sailing, scuba diving, cultural studies. **Destinations:** Baja California, Alaska, Arctic regions, Amazon, Costa Rica, South Africa, Florida coast, Mississippi islands, Egypt. **Special Emphasis:** Oceanography, cultures of coastal cities, marine mammals, reef studies. **Average Cost:** $2,300-$5,000.

85. **INTERTREK**
P.O. Box 50488
Phoenix, AZ 85046
(800) 346-4567, Fax (617) 259-1552
Types of Adventure: Trekking and touring in central Asia. **Destinations:** Asia. **Special Emphasis:** Cultural traditions, wildlife and rugged scenery. **Average Cost:** $2,350-$3,625.

86. **ISLAND PACKERS**
1867 Spinnaker Dr.
Ventura Harbor, CA 93001
(805) 642-1393
Types of Adventure: Recreation and research-oriented travel. **Destinations:** Channel Islands, California. **Special Emphasis:** Preservation of the islands, flora, fauna and marine life. **Average Cost:** $37 for a one-day trip.

87. **JAMES HENRY RIVER JOURNEYS**
P.O. Box 807
Bolinas, CA 94924
(800) 786-1830
Types of Adventure: Adventure travel in threatened areas. **Destinations:** North America and the Far East. **Special Emphasis:** Education about the effects of industry and development on the environment. **Average Cost:** $80-$3,125.

88. **JAPAN & ORIENT TOURS, INC./J&O HOLIDAYS**
3131 Camino del Rio N., Ste. 1080
San Diego, CA 92108
(619) 282-3131, (800) 377-1080
Types of Adventure: Trekking (Nepal, Burma, Thailand, New Zealand), adventure ecotravel (Papua New Guinea), whitewater rafting (Nepal), diving (South Pacific and Southeast Asia), bicycling (Nepal). **Destinations:** Orient and South Pacific. **Special Emphasis:** Custom-designed independent adventure travel with special emphasis on Nepal and Papua New Guinea. **Average Cost:** $1,500-$5,000/person.

89. **JOURNEY TO THE EAST, INC.**
P.O. Box 1334
Flushing, NY 11352-1334

(718) 358-4034, (800) 366-4034
Types of Adventure: Hiking, bicycling and bus trips that emphasize contact with the indigenous cultures of the various ethnic groups in China, Tibet and Mongolia. **Destinations:** China, Tibet, Mongolia, Nepal and Asia. **Special Emphasis:** Opportunity to meet indigenous people and participate in their festivals. Remote locations, small groups, photography, scenery and affordable prices. **Average Cost:** Varies.

90. **JOURNEYS**
4011 Jackson Rd.
Ann Arbor, MI 48103
(800) 255-8735
Types of Adventure: Trekking, hiking, wildlife safaris, family trips, worldwide nature and culture explorations. **Destinations:** Nepal, Ladakh, Sikkim, Papua New Guinea, Australia, Indonesia, Japan, Thailand, Bhutan, Kenya, Tanzania, Botswana, Namibia, Zimbabwe, Madagascar, Hawaii, Norway, Latin America. **Special Emphasis:** Small groups, expert local guides, cross-cultural emphasis, first-class service, remote destinations. **Average Cost:** $80-$150/day.

91. **JOURNEYS EAST**
2443 Fillmore St., #289
San Francisco, CA 94115
(510) 601-1677, (415) 647-9565
Types of Adventure: Adventure and backcountry travel. **Destinations:** Japan. **Special Emphasis:** Participatory cross-cultural experience. **Average Cost:** $3,085-$3,725.

92. **KER AND DOWNEY SAFARIS**
13201 NW Freeway, Ste. 850
Houston, TX 77040
(800) 423-4236
Types of Adventure: Photographic safaris, trekking. **Destinations:** Tanzania, Botswana, Nepal. **Special Emphasis:** Conservation, ecology, minimum impact. **Average Cost:** $720-$3,615.

93. **LANDER LLAMA COMPANY—WILDERNESS PACK TRIPS**
327 Washington St., BW
Lander, WY 82520
(307) 332-5624
Types of Adventure: Wilderness llama-pack trips into the pristine wilderness of western Wyoming. **Destinations:** Wind Rivers, Absarokas Popo Agie Wilderness, Fitzpatrick Wilderness, Washakie Wilderness. **Special Emphasis:** Low-impact, customized trips. Glacial cirques. Goldens. **Average Cost:** $110/day; $87/day group rate.

94. **LITTLE ST. SIMONS ISLAND**
P.O. Box 1078
St. Simons, GA 31522
(912) 638-7472
Types of Adventure: Country inn on private

barrier island wilderness preserve. **Destinations:** Little St. Simons Island. **Special Emphasis:** Interpretive natural history programs, birding, barrier island ecology, canoeing, fishing, horseback riding. **Average Cost:** $175/person, all-inclusive.

95. **MAINE SPORT OUTFITTERS**
P.O. Box 956
Rockport, ME 04856
(207) 236-8797, (800) 722-0826 in Maine
Types of Adventure: Sea kayaking, wilderness canoe trips, outdoor school, fishing trips, island ecology, seascape photography, youth day camps. **Destinations:** All of Maine's coastal waters, islands and inland waterways. **Special Emphasis:** Sea kayaking instruction and tours, island camping and exploring, natural history topics. **Average Cost:** $75 for day trips, $100/day for extended trips, all-inclusive.

96. **MARINE ADVENTURE SAILING TOURS**
945 Fritz Cove Rd.
Juneau, AK 99801
(907) 789-0919
Types of Adventure: Custom-designed trips for groups up to 6 in Glacier Bay and trips from southeast Alaska to Puget Sound. **Destinations:** Southeast Alaska, British Columbia, Puget Sound. **Special Emphasis:** Glacier Bay, West Chichagof Wilderness Area, Tracy Arm Wilderness Area. Family trips. **Average Cost:** $125-$500/day.

97. **MAYA EXPEDITIONS**
15; Calle 1-91 Zona 10
Edificio Tauro #104
Guatemala City, Guatemala
(5022) 374666, Fax (5022) 321836
Types of Adventure: Low-impact ecotourism, guide training courses for Central Americans and a student exchange program. **Destinations:** Guatemala. **Special Emphasis:** Rafting. **Average Cost:** Varies.

98. **MICATO SAFARIS**
15 W. 26th St.
New York, NY 10010
(212) 545-7111
Types of Adventure: Photographic safaris to East Africa, tours of India. **Destinations:** Kenya, Tanzania, India. **Special Emphasis:** Learning about the culture and wildlife of the country visited. **Average Cost:** $2,250-$3,985.

99. **MINGAN ISLAND CETACEAN STUDY/ RESEARCH EXPEDITIONS**
285 Green St.
St. Lambert, QC, J4P IT3, Canada
(514) 465-9176
Types of Adventure: Research studies of marine mammals, emphasis on baleen whales. **Destinations:** Mingan Islands, Gulf of St. Lawrence, Quebec and Baja California, Mexico. **Special Emphasis:** Behavioral studies of blue, fin, humpback and minke whales. **Average Cost:** $139/day.

100. **MOUNTAIN TRAIL HORSE CENTER INC.**
RD. 2, Box 53
Wellsboro, PA 16901
(717) 376-5561
Types of Adventure: Overnight horseback camping trips in Pennsylvania wilderness, ride and cross-country skiing combination trips, ride and raft combinations. **Destinations:** Pennsylvania Grand Canyon country and state forest in north-central Pennsylvania. **Special Emphasis:** Adventure trips with personalized service. Minimum-impact camping, wildlife sightings and information, fall foliage, mountain laurel, rides to country inn. **Average Cost:** $60-$125/day.

101. **MOUNTAIN TRAVEL/SOBEK, THE ADVENTURE COMPANY**
6420 Fairmount Ave.
El Cerrito, CA 94530-3606
(510) 527-8100, (800) 227-2384
Types of Adventure: Environmental-education trips exploring the hottest environmental issues in remote areas of the world. Tours operated with noted environmental organizations, including Natural Resources Defense Council, Rainforest Action Network and National Wildlife Federation. **Destinations:** Worldwide, including the Amazon rainforest, Lake Baikal (former USSR), Alaska, Africa, Poland, Hawaii, Antarctic, Indonesia, Galápagos Islands. **Special Emphasis:** Interaction with indigenous cultures such as Xiavane tribe in Brazil, working with indigenous people to develop ecotourism programs in their countries. **Average Cost:** $2,000-$3,500.

102. **NANTAHALA OUTDOOR CENTER**
41 Hwy. 19 W.
Bryson City, NC 28713
(704) 488-2175, Fax (704) 488-2498
Types of Adventure: Whitewater kayaking, canoeing and rafting trips. **Destinations:** US, Central and South America, Nepal and Malaysia. **Special Emphasis:** Preservation of natural river habitats. **Average Cost:** $15-$2,000.

103. **NATIONAL OUTDOOR LEADERSHIP SCHOOL**
P.O. Box AA, Dept. ET
Lander, WY 82520
(307) 332-6973, Fax (307) 332-3631
Types of Adventure: Outdoor-education program that teaches leadership and wilderness skills. **Destinations:** North America, Mexico, Kenya, Chile. **Special Emphasis:** A core curriculum including safety, minimal impact, outdoor and environmental studies, ecology and land management. **Average Cost:** $950 (2 weeks), $6,750 (1 semester), $2,300 (1 month).

104. **NATURAL HABITAT WILDLIFE ADVENTURES**
1 Sussex Station
Sussex, NJ 07461
(800) 543-8917
Types of Adventure: Visiting animals in their own habitat. **Destinations:** Canada, Bahamas, Alaska, Central America, Baja, Mexico, Galápagos Islands. **Special Emphasis:** Replacing hunting revenue with tourist dollars. **Average Cost:** $1,495-$5,495.

105. **NATURE EXPEDITIONS INTERNATIONAL**
P.O. Box 11496
Eugene, OR 97440
(503) 484-6529, (800) 869-0689
Types of Adventure: Natural history and cultural expeditions led by professional educators with advanced degrees in science. **Destinations:** Africa (Kenya, Tanzania, Rwanda), Asia (India, Nepal), Oceania (Australia, New Zealand), South America (Galápagos, Amazon), Central America (Costa Rica), North America (Alaska, Hawaii). **Special Emphasis:** Education-oriented approach to nature study and cultural interactions. **Average Cost:** $2,190 for a 16-day trip.

106. **NEW AND GAULEY RIVER ADVENTURES**
Box 44
Lansing, WV 25862
(304) 574-3008, (800) SKY-RAFT
Types of Adventure: Whitewater rafting, scenic float trips, mountain biking, rock climbing and horseback trips. **Destinations:** West Virginia, New River Gorge National River, Gauley and Meadow National Recreation Area. **Special Emphasis:** Environmentally sensitive company offering guided whitewater rafting along with local history, geology of West Virginia. Special exotic trips to South America. **Average Cost:** $45-$800.

107. **NEW ZEALAND ADVENTURES**
11701 Meridian N.
Seattle, WA 98133
(206) 364-0160, Fax (206) 368-8941
Types of Adventure: Sea kayaking along the shores of New Zealand. **Destinations:** New Zealand. **Special Emphasis:** The flora and fauna of the South Pacific. **Average Cost:** $750-$1,195.

108. **NICHOLS EXPEDITIONS**
497 N. Main
Moab, UT 84532
(801) 259-7882, (800) 635-1792
Types of Adventure: Mountain biking, trekking, sea kayaking, backpacking, rafting. **Destinations:** Canyonlands National Park, Grand Canyon, Salmon River and hot springs (Idaho), Alaska, Baja, Thailand. **Special Emphasis:** Low-impact camping and wilderness skills, environmental ecology and specific, sport-related skills taught. **Average Cost:** $50-$100/day.

109. **NORTHWEST VOYAGEURS LTD.**
#1 Voyageurs Blvd., Ste. A.
Lucile, ID 83542
(208) 628-3021, (800) 727-9977
Types of Adventure: Mountain biking tours, trekking, whitewater rafting, photography workshops, horseback trips, canyon "cleanup" trips. **Destinations:** Hells Canyon of the Snake River, The Salmon—"River of No Return," 7 Devils Wilderness Area, Ladakh, Garwhall Himalayas. **Special Emphasis:** Education, minimal impact, learning new skills, teaching new travel techniques, spring "high water" trips. **Average Cost:** $150/person/day.

110. **OSPREY EXPEDITIONS**
P.O. Box 209
Denali National Park, AK 99755
(907) 683-2734
Types of Adventure: Deluxe wilderness raft expeditions, 2-12-day trips in Alaska. **Destinations:** Copper River, Chitina River, Talkeetna River, Yanert Fork, South Fork Kuskoywim, Tazlina River, Nenana River. **Special Emphasis:** Wilderness, wildlife, minimal-impact camping, small group, personal attention, first-class service. **Average Cost:** $175/person/day, all-inclusive.

111. **OUTDOOR DISCOVERIES**
P.O. Box 7687
Tacoma, WA 98407
(206) 759-6555
Types of Adventure: Sea kayaking and climbing excursions. **Destinations:** Western and southwestern US, Scotland. **Special Emphasis:** Minimal-impact backcountry travel with cultural and environmental education. **Average Cost:** $1,500.

112. **OUTER EDGE EXPEDITIONS**
45500 Pontiac Trail, Ste. B
Wallod Lake, MI 48390
(313) 624-5140, (800) 322-5235.
Types of Adventure: One- to ten-person adventure expeditions with close wildlife encounters in an ecologically sound fashion. Kayaking among whales, swimming with pink dolphins, traveling in a dugout or on a camel, trekking with goats, mushing a dogsled. **Destinations:** Amazon jungle, Australia, British Columbia, Canadian Rockies, Great Barrier Reef, Machu Picchu, Peru, Vancouver Island. **Special Emphasis:** Very small groups, unusual and remote wilderness locations, traveling lightly with minimal environmental impact, natural history education. **Average Cost:** $90-$220/day.

113. **OVERSEAS ADVENTURE TRAVEL**
349 Broadway
Cambridge, MA 02139

(617) 876-0533, (800) 221-0814
Types of Adventure: Trekking, photo safaris, family adventure, summit climbing. **Destinations:** Tanzania, Costa Rica, Galápagos, Venezuela, Botswana, Morocco, Amazon, Egypt, Chile, Bolivia, South Africa, Zambia. **Special Emphasis:** Natural history tours, minimal-impact trips. **Average Cost:** $1,200-$3,200 for a 2-week trip.

114. PARADISE BICYCLE TOURS
P.O. Box 1726
Evergreen, CO 80439
(303) 670-1842, (800) 626-8271
Types of Adventure: Bicycling, hiking, snorkeling/diving, wildlife ecology, indigenous cultures. **Destinations:** East Africa, Kenya, Central America, Belize, Tanzania, Australia, New Zealand. **Special Emphasis:** Fully vehicle-supported trips, wildlife viewing, cultural interaction. **Average Cost:** $1,500.

115. POWDER RIDGE SKI TOURING
7124 W. Highway 30
Petersboro, UT 84325
(801) 752-9610
Types of Adventure: Backcountry skiing with overnight lodging at yurts. **Destinations:** Logan Canyon, northern Utah. **Special Emphasis:** Backcountry powder skiing with backcountry yurt lodging, winter skills instruction, guide service. **Average Cost:** $120/night for yurt rental, sleeps 6; guided trips available, cost varies.

116. R.A.I.N.
P.O. Box 4418
Seattle, WA 98104
(206) 324-7163, Fax (206) 562-1727
Types of Adventure: Sea kayaking, swimming, rafting and birdwatching with local guides. **Destinations:** Belize. **Special Emphasis:** Educational nature tours in Belize that support the Monkey Bay Wildlife Sanctuary. **Average Cost:** $1,400-$2,240.

117. RARA AVIS S.A.-RAINFOREST LODGE AND RESERVE
P.O. Box 8105-1000
San José, Costa Rica
(506) 53-0844
Types of Adventure: Guided nature walks in a private rainforest. **Destinations:** Costa Rica. **Special Emphasis:** Rainforest ecology, research and preservation issues are discussed by naturalist guides. **Average Cost:** Varies.

118. REI ADVENTURES
P.O. Box 1938
Sumner, WA 98390
(206) 891-2631, (800) 622-2236
Types of Adventure: Bicycling, kayaking, climbing, trekking and walking. **Destinations:** Asia, US, New Zealand, Australia, South America, Central America, Europe, Commonwealth of Independent States (former Soviet Union). **Special Emphasis:** Active adventure travel in small groups with hands-on experience. **Average Cost:** $1,000-$1,500.

119. RIOS TROPICALES
P.O. Box 472-1200
San José, Costa Rica
(011) 506-33-6455
Types of Adventure: Whitewater rafting and kayaking through primary and secondary forests. **Destinations:** Costa Rica. **Special Emphasis:** Water sports through remote areas, nature education. **Average Cost:** $60-$80/day.

120. RIVER TRAVEL CENTER
Box 6
Point Arena, CA 95468
(707) 882-2258
Types of Adventure: Wilderness rafting and sea kayaking. **Destinations:** Western US, Central America, western Canada and other international destinations. **Special Emphasis:** Wilderness adventure. **Average Cost:** $100/day.

121. ROCKY MOUNTAIN RIVER TOURS
P.O. Box 2552-BW
Boise, ID 83701
(208) 345-2400
Types of Adventure: Wilderness paddle rafting and kayaking. **Destinations:** Idaho's Middle Fork Salmon. **Special Emphasis:** No motorized boats, no crowds, small groups, all equipment provided. **Average Cost:** $745 for 4 days, $1,145 for 6 days.

122. SAFARI CONSULTANTS, LTD.
4N211 Locust Ave.
West Chicago, IL 60185
(708) 293-9288, (800) 762-4027
Types of Adventure: Hard and soft adventure travel throughout Africa. **Destinations:** Botswana, Namibia, Zimbabwe, Zambia, South Africa, Kenya, Tanzania. **Special Emphasis:** Firsthand exposure to learning about the effects of tourism, game management and wildlife preservation under the direction of government and public institutions. **Average Cost:** $2,200 for a 13-day safari.

123. SAFARICENTRE
Box 309
Manhattan Beach, CA 90266
(310) 546-4411
Types of Adventure: Overland safaris, photo safaris, trekking, camping, river expeditions. **Destinations:** Africa, Asia, Latin America, Oceania, North America. **Special Emphasis:** Nature and adventure safaris worldwide. **Average Cost:** $75-$200/day.

124. SAILING YACHT "STRANGER"
8145 Oak Park Rd.
Orlando, FL 32819
(407) 578-8792

Types of Adventure: Sailing, snorkeling, coral reef ecology. **Destinations:** US and Virgin Islands. **Special Emphasis:** Education, participation, relaxation, discussion of life's big questions. **Average Cost:** $80-$100/person/day, including food and drink.

125. SCHOONERS HERITAGE, ISAAC H. EVANS, LEWIS R. FRENCH AND AMERICAN EAGLE
Box 482-E
Rockland, ME 04841
(207) 594-8007, (800) 648-4544

Types of Adventure: Maine windjammer sailing vacations. **Destinations:** Coastal islands of Maine. **Special Emphasis:** Beautiful scenery, snug harbors, island exploring, great Down East cooking, exciting sailing, fun, informal, lobster cookouts, great memories. **Average Cost:** $295-$615/person for 3- and 6-day cruises, all-inclusive.

126. SEA TREK
P.O. Box 561
Woodacre, CA 94973
(415) 488-1000

Types of Adventure: Sea kayaking expeditions. **Destinations:** Baja California, Sea of Cortez. **Special Emphasis:** Expert kayak guiding in remote wilderness, natural history experiences. **Average Cost:** $800 for 7 days.

127. SIERRA CLUB OUTINGS
730 Polk St.
San Francisco, CA 94109
(415) 923-5522

Types of Adventure: Backpacking, bicycling, skiing, trekking, canoeing, pack-supported and base camp trips. **Destinations:** US including Alaska and Hawaii, Africa, Asia, Europe, Latin America, Pacific Basin. **Special Emphasis:** Adventure travel and service trips during which participants build and repair trails in national parks and forests. **Average Cost:** $335 for backpack trips, $225 for service trips, $2,500 for international trips.

128. SILVER CLOUD EXPEDITIONS
P.O. Box 1006-B
Salmon, ID 83467
(208) 756-6215

Types of Adventure: Wilderness whitewater trips, wilderness steelhead fishing trips. **Destinations:** River of No Return Wilderness trips, Salmon River. **Special Emphasis:** Client participation, natural history, evening campfire programs, wilderness ethics, low-impact camping. **Average Cost:** $600-$700 for summer trips, $900-$1,000 for steelhead fishing trips.

129. SLICKROCK ADVENTURES
P.O. Box 1400
Moab, UT 84532
(801) 259-6996

Types of Adventure: Belize sea kayaking, Mexico river trips. **Destinations:** Belize barrier reef, Jatate and Usumacinta rivers in Lacandon rainforest, Mexico. **Special Emphasis:** Wilderness setting, Local cultures, unique geography, tropical rainforest. **Average Cost:** $120/day.

130. SOUTHWIND ADVENTURES, INC.
P.O. Box 621057
Littleton, CO 80162-1057
(303) 972-0701

Types of Adventure: Minimum-impact trekking, camping, rafting, climbing, jungle expeditions, cultural exchanges, mountain biking, bird and wildlife viewing. **Destinations:** South America, Venezuela, Ecuador, Peru, Bolivia, Argentina, Chile. **Special Emphasis:** Experiencing and protecting nature and native cultures. **Average Cost:** $1,200-$1,700 for 13 to 17 days, varies according to destination and length of trip.

131. SPECIAL EXPEDITIONS, INC.
720 Fifth Ave.
New York, NY 10019
(212) 765-7740, (800) 762-0003

Types of Adventure: Adventure voyages aboard comfortable expedition ships accommodating less than 80 passengers, plus tented safaris to Africa and other land expeditions to far-flung destinations worldwide. Expeditions are environmentally responsible, educating travelers about the places visited with an expert staff of naturalists, historians, geologists and archaeologists. **Destinations:** Alaska, Baja California, Columbia and Snake rivers, US Southwest, Caribbean, Belize, Costa Rica, Amazon, Orinoco River, Galápagos, British Isles, Arctic Norway, Western Europe, Baltics, Russia, Egypt, Morocco, East Africa, Australia, Papua New Guinea. **Special Emphasis:** Focus is on nature's wonders, wildlife and diverse cultures. **Average Cost:** $275-$550/day.

132. SUE'S SAFARIS, INC.
P.O. Box 2171
Rancho Palos Verdes, CA 90274-8171
(310) 541-2011, (800) 541-2011

Types of Adventure: Safaris to Africa. **Destinations:** Kenya, Tanzania, Rwanda, South Africa, Namibia, Botswana, Zimbabwe, Madagascar, Seychelles, Egypt. **Special Emphasis:** Observing wildlife and plants in native habitat, native tribes, minimal impact on environment. **Average Cost:** $6,000-$7,000.

133. TAMU SAFARIS
P.O. Box 247
West Chesterfield, NH 03466

(800) 766-9199, (802) 257-2607

Types of Adventure: Customized natural history and cultural tours to off-the-beaten-track areas where comfortable camping and wilderness lodges are combined with exciting cross-cultural activities. **Destinations:** Botswana, Kenya, Zimbabwe, Tanzania, Madagascar, Seychelles Islands. **Special Emphasis:** Ecologically and socially responsible travel for individuals and groups, including wildlife and birdwatching tours, relaxed safari camps, wilderness lodges, canoe trips, photography, cross-cultural explorations, family trips, personalized service, reasonable costs. **Average Cost:** Varies.

134. TAYLOR-CASSLING, LTD., TRAVEL PROFESSIONALS
4880 River Bend Rd.
Boulder, CO 80301
(303) 442-8585

Types of Adventure: Trekking, photo safaris, rafting, bicycling, climbing. **Destinations:** Worldwide. **Special Emphasis:** Complete planning, inclusive travel, minimum impact. **Average Cost:** $1,250/person, inclusive.

135. TERRA ADVENTURES, INC.
70-15 Nansen St.
Forest Hills, NY 11375
(718) 520-1845, Fax (718) 575-8316

Types of Adventure: Minimal-impact trekking. **Destinations:** Central and South America. **Special Emphasis:** Preservation of tourism destinations and education about tourist impact. **Average Cost:** $300-$2,000.

136. TIMBERLINE BICYCLE TOURS
7975 E. Harvard, #J
Denver, CO 80231
(303) 759-3804

Types of Adventure: Bicycle tours (on-road and off-road), biking/hiking, biking/rafting, biking/canoeing, biking/kayaking combinations. **Destinations:** Glacier, Yellowstone, Bryce, Zion, Grand Canyon, Canyonlands, Lassen and Olympic national parks, Colorado, Canadian Rockies, Puget Sound, Great North Woods. **Special Emphasis:** Fully supported, inn-to-inn tours of 38 national parks and monuments in US and Canada. **Average Cost:** $595 for 5-day tour, $795 for 7-day tour, $925 for 9-day tour.

137. TOFINO EXPEDITIONS
114-1857 W. 4th Ave.
Vancouver, BC, V6J 1M4, Canada
(604) 737-2030, Fax (604) 737-7348

Types of Adventure: Sea kayaking. **Destinations:** Canada, Mexico, Alaska, western North America. **Special Emphasis:** Broad spectrum of environmental interpretation, accessible to beginning sea kayakers, emphasis on low-impact camping techniques. **Average Cost:** $650-$1,000/week.

138. TUSCARORA CANOE OUTFITTERS
870 Gunflint Trail
Grand Marais, MN 55604
(218) 388-2221

Types of Adventure: Canoeing and camping trips into Boundary Waters Canoe Area Wilderness of Minnesota and Quetico Park of Canada. **Destinations:** Minnesota, Ontario. **Special Emphasis:** Low-impact canoeing and camping in remote areas, wildlife viewing, good fishing, few people. **Average Cost:** $200/person/day.

139. UNIVERSITY RESEARCH EXPEDITIONS PROGRAM
University of California
Berkeley, CA 94720
(510) 642-6586

Types of Adventure: Joining scientists in researching crucial problems that threaten life on Earth. **Destinations:** Worldwide. **Special Emphasis:** Demystifying the scientific method and involving as many people as possible in the rewards of discovery. **Average Cost:** $1,400.

140. VICTOR EMANUEL NATURE TOURS, INC.
P.O. Box 33008
Austin, TX 78764
(512) 328-5221, (800) 328-VENT

Types of Adventure: Birding tours, natural history tours, photo tours. **Destinations:** Worldwide. **Special Emphasis:** Birds, nature. **Average Cost:** $150/day.

141. VOYAGERS INTERNATIONAL
P.O. Box 915
Ithaca, NY 14851
(607) 257-3091, (800) 633-0299

Types of Adventure: Natural history, nature photography, birding tours led by outstanding naturalists and professional photographers. **Destinations:** Worldwide, Africa, Galápagos, Costa Rica, New Zealand, Australia, Ireland, Hawaii, Nepal, Antarctica, Alaska, Indonesia, India, Belize. **Special Emphasis:** Outdoor photography, birding, natural history, small groups, low-impact trips. **Average Cost:** $2,000-$5,000.

142. WHITE MAGIC UNLIMITED
P.O. Box 5506
Mill Valley, CA 94942
(415) 381-8889, (800) 869-9874

Types of Adventure: Rafting, trekking, diving, jungle safaris, mountain biking, whale watching, cultural tours, natural history, archaeology. **Destinations:** Alaska, Grand Canyon, Middle and Main Salmon rivers, Mexico, Guatemala, Costa Rica, Chile, Belize, Nepal, India, Thailand, Turkey, Albania, US Southwest, California. **Special Emphasis:** Actualizing dreams on Earth. **Average Cost:** $500-$2,500.

143. WHITEWATER SPECIALTY
N3894 Highway 55
White Lake, WI 54491
(715) 882-5400
Types of Adventure: Canoe and kayak paddling school. **Destinations:** Central US. **Special Emphasis:** Professional instruction in canoeing and kayaking. **Average Cost:** $275.

144. WILDERNESS ALASKA
P.O. Box 113063
Anchorage, AK 99511
(907) 345-3567
Types of Adventure: Scheduled and custom backpacking, rafting, base camp trips. **Destinations:** Brooks Range, Alaska, Arctic National Wildlife Refuge. **Special Emphasis:** Small groups in remote wilderness unfolding the unique natural history of the Brooks Range. **Average Cost:** $1,800.

145. WILDERNESS RIVER OUTFITTERS AND TRAIL EXPEDITIONS, INC.
P.O. Box 871
Salmon, ID 83467
(208) 756-3959
Types of Adventure: Whitewater rafting, backpacking, biking, ski tours. **Destinations:** Idaho, Montana, Alaska, New Zealand, Chile. **Special Emphasis:** Wilderness adventure. **Average Cost:** $120/day.

146. WILDERNESS SOUTHEAST
711 Sandtown Rd.
Savannah, GA 31410
(912) 897-5108
Types of Adventure: Wilderness camping, canoeing/kayaking, hiking, backpacking, snorkeling, natural history education. **Destinations:** Okefenokee Swamp, Everglades, Great Smoky Mountains, Cumberland Island, Florida springs, British Virgin Islands, Bahamas, Belize, Costa Rica, Amazon Basin and the Pantanal. **Special Emphasis:** Natural history, ecology. **Average Cost:** $70-$100/day.

147. WILDERNESS TRAVEL
801 Allston Way
Berkeley, CA 94710
(510) 548-0420, (800) 368-2794
Types of Adventure: Adventure travel, natural history, wildlife and cultural expeditions around the world, 100 different itineraries, over 300 departures, hiking, trekking, adventure cruises, jungle and safari trips. **Destinations:** South America, Galápagos, Asia, Pacific and Pacific Rim, Africa, Europe. **Special Emphasis:** Small groups with expert leaders, unrivaled customer service, innovative itineraries. **Average Cost:** $2,000/person/land cost.

148. WILDLAND ADVENTURES
3516 NE 155th St.
Seattle, WA 98155
(206) 365-0686, (800) 345-4453
Fax (206) 363-6615
Types of Adventure: Wildlife safaris, trekking, jungle expeditions, hiking, conservation, worldwide nature and culture explorations to Earth's wild places and exotic cultures since 1978. Small groups, uncommon itineraries, personal cross-cultural interaction, pre-trip planning assistance and personal attention. **Destinations:** Andes, Amazon, Himalayas, Africa, Costa Rica, Belize, Mexico, Alaska, Turkey. **Special Emphasis:** Trips support conservation and community development projects through our nonprofit, affiliated conservation organization, The Earth Preservation Fund. **Average Cost:** $1,500/land cost.

149. WOODSWOMEN
25 W. Diamond Lake Rd.
Minneapolis, MN 55419
(612) 822-3809, (800) 279-0555
Types of Adventure: Trekking Nepal with native women sherpanis, exploring Costa Rica with local naturalist, cruising the Galápagos Islands. **Destinations:** Nepal, Africa, Ecuador, Switzerland, Mexico, Costa Rica, Ireland, US, New Zealand. **Special Emphasis:** Women's adventure travel for women of all ages, skill levels and backgrounds. **Average Cost:** $765-$5,695.

150. ZEGRAHM EXPEDITIONS
1414 Dexter Ave. N., Ste. 327
Seattle, WA 98109
(206) 285-4000
Types of Adventure: Cultural expeditions, diving, expedition cruises, natural history expeditions, photography expeditions. **Destinations:** Antarctica, Arctic, Asmat, Galápagos, Amazon, Cambodia, Vietnam, Laos, Botswana. **Special Emphasis:** Small groups with expert leadership. **Average Cost:** $2,000-$8,000.

The Best Environmental Magazine In The World!

Buzzworm is an old southwest term for rattlesnake. When a rattlesnake "buzzes," it is a warning that must be heeded. BUZZWORM is also the name of an exciting magazine that warns of serious threats to our Earth.

If you're concerned about the environment and want to know more about it, BUZZWORM is for you. Through insightful writing and spectacular full-color photography, BUZZWORM explores the hows and whys of sustaining our fragile planet. As a "consumer guide" to the environment, BUZZWORM also features in-depth listings of volunteer and job opportunities, and ecological adventure travel.

WINNER:

- *Magazine Week's* Editorial Excellence Award
- Maggie Award for Best Special Interest Magazine
- World Hunger Media Award for Best Periodical

Not affiliated with any environmental organization, BUZZWORM is acclaimed as the first truly independent publication reporting on environmental issues.

If you like *Earth Journal*, you will love BUZZWORM. Examine this exceptional magazine for yourself, without obligation. Simply fill out and return the coupon below (or a photocopy) to receive a no-risk FREE ISSUE.

YES! I accept your offer. Send me my risk-free issue of BUZZWORM. If I like it I will pay just $18.00 for a full year (six big bimonthly issues, including my free issue) and save $6.00, 25% off the regular cover price. If I choose not to subscribe, I'll write "cancel" on the bill, owe nothing, and keep the issue—FREE!

Name _____

Address _____

City/State/Zip _____

793TRX

Send to: BUZZWORM, P.O. Box 6853, Syracuse, NY 13217-7930

GUARANTEE: We are dedicated to making BUZZWORM the best environmental magazine in the world. If you are ever dissatisfied with BUZZWORM, for any reason, just let us know. You'll get a prompt and unquestioned refund on all unmailed issues.